Labor and the Locavore

Labor and the Locavore

THE MAKING OF A COMPREHENSIVE FOOD ETHIC

Margaret Gray

UNIVERSITY OF CALIFORNIA PRESS

BERKELEY LOS ANGELES LONDON

University of California Press, one of the most distinguished university presses in the United States, enriches lives around the world by advancing scholarship in the humanities, social sciences, and natural sciences. Its activities are supported by the UC Press Foundation and by philanthropic contributions from individuals and institutions. For more information, visit www.ucpress.edu.

University of California Press
Berkeley and Los Angeles, California

University of California Press, Ltd.
London, England

Library of Congress Cataloging-in-Publication Data

Gray, Margaret.
    Labor and the locavore : the making of a comprehensive food ethic / Margaret Gray.
        p.   cm.
    Includes bibliographical references and index.
    ISBN 978-0-520-27667-3 (cloth, alk. paper)
    ISBN 978-0-520-27669-7 (pbk., alk. paper)
        1. Agricultural laborers—Abuse of—Hudson River Valley (N.Y. and N.J.)   2. Agricultural laborers—Employment—Hudson River Valley (N.Y. and N.J.)   3. Labor policy—Hudson River Valley (N.Y. and N.J.)
    I. Title.
    HD1527.N7G73   2013
    331.2'97473—dc23   2013011414

Manufactured in the United States of America

22   21   20   19   18   17   16   15   14   13
10   9   8   7   6   5   4   3   2   1

In keeping with its commitment to support environmentally responsible and sustainable printing practices, UC Press has printed this book on Cascades Enviro 100, a 100% post consumer waste, recycled, de-inked fiber. FSC recycled certified and processed chlorine free. It is acid free, Ecologo certified, and manufactured by BioGas energy.

CONTENTS

ILLUSTRATIONS

TABLE

MAP

FIGURES

ACKNOWLEDGMENTS

This book would not have been possible without the cooperation of my many interviewees, the vast majority of whom were assured confidentiality and whom I cannot thank by name. I am deeply grateful to the farmworkers who welcomed me into their homes and spoke to me about their lives. I also want to thank the growers who were willing to share their stories and opinions on farm labor. In addition, I am greatly indebted to farmworker advocates and service providers who let me join their ranks; in particular, thank you to Richard Witt and the late Jim Schmidt.

A grant from the ILGWU 21st Century Heritage Fund allowed me to conduct interviews with Hudson Valley farmworkers; further grants from the Richard Styskal Dissertation Fellowship and Adelphi University supported my research and writing. Additional research and writing was conducted while I was a Rockefeller postdoctoral fellow at Stony Brook University's Latin American and Caribbean Studies Center, where I met Paul Gootenberg, Javier Auyero, Eileen Otis, Christina Ewig, and Luis Reygades, all of whom offered valuable advice on my project.

Frances Fox Piven has been a staunch advocate of my work and helped me develop a strong sense of my methodology, data analysis, and writing. She offered invaluable advice and let me disagree with her whenever I wanted. Other mentors include Andrew Polsky, Juan Flores, John Mollenkopf, Robert Smith, Lenny Markovitz, and Ken Erickson. Colleagues who offered important insights on earlier versions of these chapters include Penny Lewis, Glen Bibler, Arielle Goldberg, Rose Muzio, Kevin Ozgercin, Miriam Jimenez, Carey Powers, Lorna Mason, Victoria Allen, Rich Meager, and Lori Minnite.

My fieldwork with farmworkers was conducted in collaboration with the Bard College Migrant Labor Project, and I would like to extend my thanks for administrative support from Bard's Institute for International Liberal Education and Human Rights Project, especially Tom Keenan, Susan Gillespie, Melanie Nicholson, and Danielle Riou. Bard College interns who were vital to the success of this project include those who helped me with farmworkers interviews (Jessica Hankey, Anna Mojallali, Katie Ray, Owen Thompson, and especially Diana Vasquez and Betsaida Alcantara); those who offered administrative help (Kyle Jastor, Sarah Gibbons, and Nikkya Martin); and those who did transcription and translation (Adrian Masters, Zoe Elizabeth Noyes, and Katherine Del Salto Calderon).

Adelphi University students tendered key assistance. Pedro Hipolito-Albarra did transcription and translation; research and editing support was provided by Mahnoor Misbah, Martin Vladimirov, Trish Hardgrove, Robert Danziger, Joe Conte, and Elizabeth Age. A special thank you goes to Alexa Savino, who tirelessly helped me in the last year of this project. Additional interns who helped with interviewing and transcribing included Francisca Castellanos, Marcia Espinosa, Doris Diaz-Kelly, Daniel Bleeks, and Bill Valente.

My gratitude extends to Courtney Dudley, Celia Roberts, and Mona Coade-Wingate for allowing me to use their photographs in this book and to Monique Razzouk for designing the map.

In the writing process, I was fortunate to have the insight of friends and colleagues. Thanks to Janice Fine, Cindy Hahamovitch, Mary Summers, Ross Cheit, Grant Farrad, Shareen Hertel, Roger Waldinger, Manny Ness, William McDermott, Federico Chicci, Ismael Garcia, Fran Benson, Julianne Rana, Herb Engman, and Jordan Wells for their comments. My understanding of labor relations was shaped by Charlie Kernaghan and Barbara Briggs, to whom I am grateful. My colleagues at Adelphi have been extraordinarily supportive. I thank Regina Axelrod, Katie Laatikainen, Traci Levy, and Hugh Wilson in addition to Dawn Kelleher, Mary Cortina, and Gayle Insler. Thank you also to Adelphi's Research Support Group, particularly Jonathan Hiller and Courtney Weida.

My family—and especially my mother, Ann Gray—has been extremely supportive of my scholarly endeavors. To friends Miguel Angel Pimentel and Peri Lyons, thank you for your insight and always knowing the right thing to say. My writing group, BAWG, offered stellar advice and moral support. I thank Krishnendu Ray, Toral Gajarawala, Thuy Linh Tu, Jessamyn Hatcher, and Julie Elman. Special gratitude goes to Toni Pole and Emma Kreyche, who provided intellectual and emotional support every step of the way. I am so appreciative of their advice. I am extremely grateful to my superego, Dana

Polan, who helped steer the direction of this book, read several versions of the manuscript, and was ever ready to ask the right questions at the right time over his expertly prepared locavore meals.

I owe Kate Marshall, my editor, *mil gracias* and a New York City cocktail for her patience and for ushering this book through its final stages. My daughters Zola and Stella offered enthusiastic nagging and much-needed respite from this project with their good humor and charm. Finally, I extend my deepest appreciation to my partner, Andrew Ross, who knew when to step forward and when to stand down, supported me in every aspect of this project, and kept laughter in the household.

I would like to dedicate this book to the memory of my father, Bob Gray, and my movement comrade, Jim Schmidt.

Introduction

IS LOCAL FOOD AN ETHICAL ALTERNATIVE?

A SEPTEMBER DRIVE UP THE Taconic State Parkway or New York Thruway or a journey along one of the region's rural roads offers storybook views of bountiful red and yellow fruit adorning rows of trees. The orchards of the Hudson Valley have been a staple of the agricultural region for several centuries. The twenty-first century has seen a revival of interest in the valley's farms as consumers visit pick-your-own farms and pull over at farm stands. For local urbanites the Hudson Valley delivers; many of the goods on offer at farmers' markets originate from the region. The media storm of food advocacy has increased our exposure to farming. We have a better feel for the seasonality of local produce; glossy photos help us see how our food grows (who expected okra to point up?); farmers' struggles with the weather and pests are more fully understood; the nuances of the meaning of "free range" are no longer lost on us; and we read about the many tasks involved in farming from sunup to sundown.

What the orchards often hide and food writers tend to avoid are the stories of the farmworkers who staff these regional farms. Unheard is Marco's complaint that his sixteen-hour workdays at minimum wage are destroying his body and Gloria's desire for a seat during her long days packing apples. Although the 2000 case against upstate New York labor contractor Maria Garcia, who pleaded guilty to forced labor, received some media coverage, we usually do not get to read about how Hector was fired from his job and lost his home on the farm when he hurt his back trimming apple trees in icy conditions. Hidden are the conditions, the expectations, and the hopes of the farm labor force.

"I aimed at the public's heart, and by accident I hit it in the stomach." So went the acerbic response of Upton Sinclair to the reception of his muckraking novel *The Jungle* about workers in Chicago's slaughterhouses.[1] The book's profound impact ushered in a round of legislative protections—

including the passage of the 1906 Pure Food and Drug Act and the 1906 Meat Inspection Act—along with a durable consensus about the need for consumer regulation and advocacy. However, neither public advocates nor policy makers responded in kind to Sinclair's depiction of the low-wage immigrant work force and their deplorable workplace conditions. In much the same way, today's burgeoning alternative food movement is in danger of repeating this injustice. Food movement advocates and consumers, driven to forge alternatives to industrial agribusiness, have neglected the labor economy that underpins "local" food production. Consequently, the injunction to "buy local" promotes public health at the expense of protecting the well-being of the farmworkers who grow and harvest the much-coveted produce on regional farms.

The U.S. public has not been reluctant to recognize the exploitation of immigrant farmworkers on factory farms, which are part of the industrial, commodity food system. But the resurgence of interest in healthy food and sustainable agriculture among academic and popular writers has overlooked the role of hired labor in smaller-scale agrifood production. Instead, it has been borne along by an idealized, agrarian vision of soil-and-toil harmony on family farms.[2] In a similar vein, it was once commonly assumed that organic fruits and vegetables were produced under more ethical conditions, until journalistic exposés and scholarly scrutiny demystified the category of "organic" by revealing how it was co-opted by big agribusiness, which influenced the weakening of federal organic standards. The "organic" label on a product no longer carries the automatic imprimatur of morally superior food. Despite the veneer of ethical production, it remains the case that local or small agricultural producers are driven by market dictates and regulatory norms that render their approach to labor relations more or less undistinguishable from those of larger, commodity-oriented, industrial farms.[3]

Localism trades on the ideals of agrarianism and the self-reliant yeomanry, ideals that have written out of their record the role of hired labor. In correcting the record, and making a case for expanding the ethical reach of food justice, this book explicates the hidden costs of agrarianist dogma and the human consequences flowing from policy makers' neglect of these costs. Food writers have promoted locavore diets as wholesome and righteous alternatives to the capitalist-industrial food system, and many small farmers have seen their businesses thrive on the back of this new consumer trend.

In the public mind, eating locally resonates with eighteenth-century republican ideals about self-sufficiency that hark back to the nation's origins. Yet countering the influence of this Jeffersonian romanticism is the growing public awareness, stimulated by heightened scrutiny of immigration, of the fact that small farms, like their factory farm counterparts, are largely staffed by noncitizen, immigrant laborers, and specifically undocumented workers and foreign guest workers. The insourcing of this cheap immigrant labor, a longstanding practice in large farming states and metropolitan areas, is now widespread in smaller farming states, as well as most service industries and in a range of suburban and smaller urban locations. Yet the prevailing mentality within the alternative food movement has not absorbed this reality. Promoting ethical consumption and demanding a shift to sustainable and just agriculture rarely include a call for justice for farmworkers. Food advocates and their organizations display a tendency to conflate *local, alternative, sustainable,* and *fair* as a compendium of virtues against the factory farm that they so vigorously demonize. Yet this equation discourages close scrutiny of the labor dynamics by which small farms maintain their operations.

Following the logic espoused by the food movement, the moral scrutiny of food production should focus in part on the livelihoods of those who labor in the fields. In other words, for food politics to truly promote public health, improving workers' conditions should be considered as important as protecting watersheds or shielding animals from pain and stress. With so much laudatory attention heaped on the small producers, their employees, by right, should be afforded the same high moral estimate. In addition, sustainability in all areas of life is increasingly recognized as an inherently valuable goal. Ethical eaters should be concerned if their own health and livelihoods are advanced at the expense of others. Exposés of the high cost of cheap food, which count the external environmental costs of industrial food production, do a disservice to readers by neglecting the farm work force.[4] The same principle should be applied to the local farm ecosystem. A better understanding of immigrant workers' role as the mainstay of U.S. agriculture is necessary if the new generation of ethical farming is going to provide sustainable jobs.

Promoting local food justice in terms of environmental protection, animal welfare, and saving small farms from the auctioneer or the bulldozer is an admirable goal for those interested in an alternative to the capitalist-industrial food system. But if a food ethic that values environmental, economic, and social goals is to extend to all those involved in production, then it must entail active support for workers' rights with a view to improving the livelihoods of the laborers. The research presented in this book—about the labor economy of regional farms in the Hudson Valley region of New York State—is an effort to take on that challenge. This book questions the

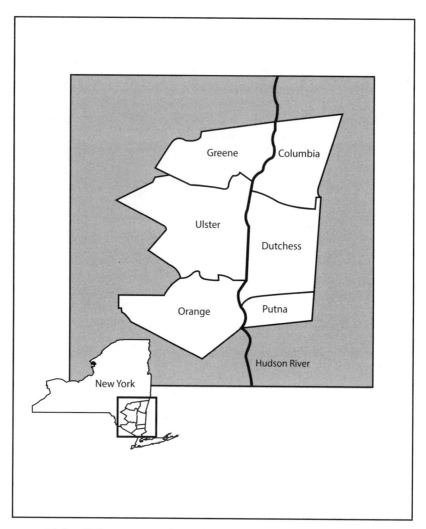

MAP 1. Hudson Valley counties in this study.

underlying power systems that shape the way we understand our food systems, large and small, and invites readers to consider what it means to embrace a more comprehensive food ethic.

．　．　．

Barry Estabrook's 2011 book *Tomatoland* presents compelling evidence about the appalling working conditions of Florida's tomato workers. Some of these conditions include forced labor, modern-day slavery, and the ruinous effects of pesticides on the health of farmworkers and their children. This book seemed to hit the heart of some food writers, who were moved to condemn the industrial food system in the hope that they could persuade food advocates to expand their understanding of sustainability to include labor concerns.[5] Estabrook follows in Sinclair's muckraking tradition, one that *New York Times* columnist Nicholas Kristof regards as the investigative journalist's most important resource in addressing social issues. According to Kristof, the use of extreme examples to bring attention to a cause is a legitimate instrument of the journalist's trade.[6] From the time I began my field research in 2000, I heard my fair share of extreme stories from and about New York's farms. Some of these include human trafficking and forced labor, and reports of child labor, sexual assault, eviction, and the preemptory firing of workers who were injured on the job. With the exception of the severe, persistent pesticide poisoning of farmworkers that Estabrook describes, I heard New York farmworkers and their advocates offer accounts of labor violations very similar to those in *Tomatoland*. But, unlike Estabrook's reporting on conditions in the industrial agrifood sector, my research focus is local farms. Nor is this book a journalistic exposé. Highlighting exceptional cases of abuse can lead us to think about poor labor practices as aberrant or isolated. In this book, however, I present the broader patterns and systematic conditions of treatment, which include meager wages, long hours of difficult manual work, lack of overtime pay, run-down housing, lack of respect, and paternalistic management practices. This has been the guiding approach for my examination of workers' everyday vulnerability and the difficulty they experience in addressing their problems.

The farmworkers I sought out and interviewed were almost all noncitizen immigrants, mostly undocumented Latinos. While Latino and undocumented workers have historically populated farming communities in other regions in the United States, they are relatively new to Hudson Valley farms. New York farms experienced a significant demographic shift in the late twentieth century, from African American and Caribbean to Latino laborers.[7] Only a fifth of farmworkers were not Latino: they were Jamaican, of whom three-quarters were foreign guest workers on special visas. Some of the overall number of workers were new arrivals, while others had been in the Hudson Valley for more than a decade. Almost a quarter were traditional migrants coming from other farms in the United States, while a third, including guest workers, returned to their home countries every year, and the remainder lived within the region year-round. An overwhelming majority (92 percent) were either undocumented or guest workers, legal statuses that render workers especially vulnerable to poor treatment and labor abuses. The majority of those I interviewed did not know their rights, and most of the Latinos spoke little to no English. As a result, they had not been "Americanized" in any substantial way, a feature that my farmer interviewees for the most part found desirable. Once they were accustomed to American ways, I was repeatedly told, they were no longer "good workers."

The new American romance with farming inspired by the food movement offers a pathway beyond the alienation of labor. Those who choose to work the land are seen as returning to nature, redeeming nutritious food, and shaping community in the lineage of the frontier settler. Yet there is no empirical correspondence here with the daily reality of a farmhand. The field hands and packing workers describe their work as arduous and dirty. It requires bending and stretching, long hours on one's feet, repetitive motions, wielding sharp tools, carrying heavy loads, and working in extremes of heat, wet, and cold. The manual work involved in planting, maintaining, and harvesting the fruit and vegetables on local farms is challenging. In climates like New York, the work is mostly seasonal, and so it cannot sustain anyone for long and affords little security. The average pay is meager by all measures; when I reinterviewed farmworkers after a period of six years, some were earning the same wage or only a slightly higher wage than when I first spoke to them. And the type of benefits guaranteed to other kinds of workers, such as sick and vacation days, health insurance, and retirement funds, were unheard of. Beyond the strenuous nature of the labor and the scant remuneration, workers took pains to describe to me the complicated relationships they had with those who hired them. Their workplace was governed by employer paternalism and the concomitant forms of labor discipline made it difficult for them to challenge substandard conditions. There were no unions to represent them and, since they sometimes numbered as few as three on a farm, there was little in the way of effective solidarity. In addition, they lacked the basic legal safeguards that most American workers enjoy, including overtime pay, a right to a day of rest, and collective bargaining protections.[8]

At a time when the American public is open to new ideas about food advocacy and sustainability, and the national spotlight is positioned on undocumented immigrants, my study aims directly at the convergence of these themes in order to focus some attention on the role of labor in the local agricultural economy. In assessing the national temper of food and immigration politics in the context of a regional agricultural labor market, I designed my research in the hope that it would yield lessons for scholars, policy makers, and activists. In the pages that follow I concentrate on answering two main questions. First, what are the factors that prevent farmworkers from making more of an effort to change their situation? Second, why have leading participants in the alternative food movement culture so far not embraced a deeper interest in farmworkers in local agricultural settings? One obvious implication that flows from these questions is that the reform energies generated by the locavore ethic have the potential to lead to improvements in the labor conditions on local farms by facilitating changes that workers and their advocates are unable to bring about on their own.

Labor and the Locavore is based on more than a decade's worth of field research—primarily in-depth interviews and participant observation—in New York's Hudson Valley and other regions in New York State. In addition to interviewing farmworkers, between 2000 and 2010 I interviewed farmers, statewide farmworker advocates, government employees, state legislators and aides, lobbyists, representatives from farmer organizations, and a range of farmworker service providers. My participant observation involved more than a hundred advocacy-related meetings within the state's farmworker justice movement, farmers' markets, and food and farm conferences, and active attendance at a wide range of public events—rallies, marches, delegations, public forums, fundraising events, community meetings, and legislative sessions and hearings. I also supplemented and compared my primary data with journalistic accounts, government information, nonprofit reports, management texts, and scholarly research on the state's agriculture industry. (For more details, see the Methodological Appendix.)

In responses to lawmakers and public concern, the New York Farm Bureau, the most prominent farmer association in the state, always insists on legally admissible evidence of "abuse" to justify the need for increased labor protections. But it is not the violations of specific labor laws so much as the institutional marginalization of agricultural workers—including the denial of basic labor rights—that reproduces their chronically unequal status within the workplace and host communities. According to existing New York State law, employees can be legally asked to work seven days a week for fifteen hours a day or more and never be paid at standard overtime rates or offered a day of rest. In addition, if workers unite to try and change their circumstances, they can be fired with legal impunity. How do these exclusions affect workers on a daily basis? To give the reader an experiential snapshot of the answer to that question, let me offer a characteristic scenario from a vegetable farm that typified the kind of workplaces I researched for this book.

On a hot summer day Jorge and Carlos arrived at the main farm building at six in the morning and learned of their tasks for the day. Crews prepared by loading empty bins into the trucks, which would return from the fields at midday and again after dark, overflowing with freshly picked vegetables. Cutting vegetables was grueling. I watched as both men performed the work on their haunches or bent over with a sharp knife. It was obvious that they had to work quickly. Daytime temperatures at this time of year regularly broke the 80-degree mark in the dusty fields, and this day was no exception. Dirt worked its way into every uncovered pore. Throats became parched, and workers did not always have time to drink enough water. On the other hand, the avoidance of liquids meant fewer bathroom breaks, which were also put off since the portable toilets in the field were often at a considerable distance.[9] During the hottest part of the day, their crew retreated to their trailers in the labor camp for an hour's lunch. A short dinner break took place in the evening with food delivered directly to the fields. The workday ended at ten, well after dark.

A few months later, when Jorge and Carlos had more free time in the evenings, I sat down with them in a trailer they shared with four others. Both told me they generally enjoyed picking vegetables, but the pace and hours often got to them. Jorge reported that the "ugly part" was being forced to endure extreme heat without many breaks, but then again, he was used to it since he'd been working in the fields since he was a child. Their compatriot Miguel summed up the crew's general acceptance of how rewards were meted out: "If you behave, there is work." It was the kind of comment about labor discipline that I heard repeated on the other farms I visited. He explained that the boss regularly came to check in about their conduct with the worker-supervisor, an employee like them who had earned trust and responsibility through hard work and loyalty, or who inherited the position through family members ("like a dynasty," Carlos remarked). He also explained that the boss relied on "the ones he works closest with," his white employees who worked as mechanics, tractor drivers, or general handymen, "to see if we are being lazy and then they call the boss who comes and reprimands us." Social life for these workers was restricted since most nights they worked until nine or later; they regularly logged ninety hours a week during the summer and early fall. On Saturday nights they went to the nearest Walmart to buy groceries and they usually ate dinner at Burger King or a Chinese restaurant (they preferred buffets). On Sundays they tried to play soccer for a few hours. Miguel had worked on the farm for twelve years and still earned only eight dollars an hour. His teenage daughter was earning the same wage in her capacity as a mother's helper, and he could not conceal his bitterness about what this comparison revealed. Though grateful, if only for the continuity of the work, Miguel worried that he was finding his situation more difficult to bear as each new year brought the same grind with no prospects of improvement.

Most of the workers I interviewed had stories similar to those of Jorge, Carlos, and Miguel. Their comments revealed the extreme social stratification of the agricultural work force, and, quite frankly, they challenged assumptions I had once made myself about the sustainability of jobs on regional farms. In 2002 the workers I interviewed had an average annual income of $8,078, a number that included not only their Hudson Valley farm labor income but also all their additional income from the year. Six years later the difference in incomes was negligible. Again and again, my interviewees explained how their desperate need for a subsistence income was the underlying incentive for their decisions and that it almost always overrode concerns about their personal well-being. They neglected health issues, took on additional work on their only day off, subsisted on canned foods because of their cost and convenience, and, for those with families in their home countries, deferred trips home for years at a time. For those who stayed year-round to try to secure nonseasonal work, the prospects were so limited that many were unemployed for most of the winter, while others secured outdoor maintenance work, such as trimming trees in the orchards or performing repairs, often in icy temperatures.

One's response to this description of how Miguel, Carlos, and Jorge live and work depends on one's perspective. Among my interviewees were employers who argued that, despite the hardships, their workers were living the American dream, bettering their circumstances as immigrants had done in the past. Others declared that their employees wanted long hours and loved their jobs. Still others insisted that, as employers facing bankruptcy at every turn, they could not afford to offer better pay. The farmworker advocates I interviewed saw these conditions as the outcome of constant labor abuse stemming from a legally precarious environment that made it impossible for workers to speak up, let alone challenge their employers. There are some truths on all sides; even the workers themselves appeared to entertain these seemingly irreconcilable points of view. In these pages I try to offer multiple standpoints—those of workers, farmers, and advocates—in order to explain why employees put up with jobs they often find so problematic.[10]

In part our responses are influenced by romantic agrarianism, a powerful concept that has shaped both the food movement and, in turn, management practices on small farms. Urban consumers in New York want local, sustainable, seasonal produce, and small farmers in the Hudson Valley cater to this new market demand. The history of how this has been possible, however, has a dark side, as the smaller-scale, locavore food industry relies heavily on an unstable, highly marginalized work force. Despite the consumer embrace of romantic agrarianism, small farmers' labor practices are akin to those seen on corporate industrial farms. A comprehensive food ethic requires an understanding of farmworkers' situations with an eye toward helping to improve them, even if this means challenging farmers.

WHY NEW YORK FARMS?

The Hudson Valley, the fabled agricultural region that lies to the north of New York City, is a particularly apposite setting for examining the absence of worker justice within the alternative food movement, as well as the many obstacles that lie in the path of workers' inclusion in the new food ethic.[11] The region's cultural identity trades on the currency of agrarian values and epitomizes precisely those farming sectors that have benefited most from the economic stimulus promised by alternative and local food movements. New York is exceeded only by California in the market value of "local" agricultural sales, and the six counties covered by my research account for more than 20 percent of those sales.[12] Moreover, it is distinctive for its concentration of small farms—92 percent of the state's farms are considered small-scale, with a gross income of less than $250,000 a year[13]—making it a robust case study for the local food movement. My geographically situated analysis of a small agricultural producer economy and its farm labor market stands as a counterpoint to the dominant agrifood research focus on the state of California.[14] The Hudson Valley is thick with food policy centers and is increasingly cited as a model local food system with sustainable relations to populations and resources.

Between the two poles of corporate factory farms of California and mom-and-pop operations that do not have hired hands at all, there is considerable variation. My interviews took place on "family" farms with as many as eighty workers and as few as three. The growers in question hired and housed mostly seasonal workers for fruit and vegetable production, although some lived on the premises year-round with or without work. I use the term "regional farms" and occasionally describe farms as "local," "small," or "smaller" to distinguish them from industrial farms in the larger farming states. Not every farm I studied provided produce directly to the consumer, but I argue that those that did not are still part of a regional labor market and its culture regardless of the destination of their products. The marked increase in research on undocumented workers in every part of the country has emphasized suburban or small urban locations, where growing Latino populations have caused profound social turbulence. By contrast, my research is a return to the rural, where the changes in the agricultural work force have less impact on existing communities.[15]

When locavores buy local food, they are securing fresher, more seasonal produce, but they are also buying an idea. Consumers are led to believe that purchasing directly from farms facilitates an intimate, trusting relationship with the farmer, and this bond seals the common understanding that the local food production process is more wholesome and morally gratifying than the industrial commodity system. This overendowment of meaning cannot but encourage locavores to set aside their concerns about the production process. Nor has such neglect been discouraged by the mercurial interest in food systems research.[16] Scholars in that field have expounded on the positive aspects of local food systems, including the economic and social benefits, the sense of justice and community facilitated by face-to-face interactions with food producers, and the civic engagement and democracy promoted by alternative agrisystems. However, an overwhelming share of the focus has been dedicated to the environment and small farmers while the security and well-being of farmworkers has been sidelined. In addition, popular writers have oversimplified the economy of alternative agriculture while glorifying the ethos of family farming.[17] Consider, for example, that in an important review essay on food studies and food movement literature, leading food studies scholar Marion Nestle describes Holly George-Warren's *Farm Aid: A Song for America*, about the struggle of small farmers, as the premier example of writing on "good, clean, fair food."[18] While the review essay credits Eric Schlosser for raising awareness about factory farm labor concerns, Nestle's overriding priority when it comes to significant social justice issues is to promote the livelihoods of small farmers. This standpoint encourages the public perception that there are no serious labor problems in agricultural economies typified by family farms.

The combined testimony of scholars, food writers, and journalists has made the public case for the improved conditions for animals and food sustainability on smaller, local farms. Those who seek out such improvements are engaging in what scholars call moral or virtuous consumption.[19] The time is ripe to introduce the consideration of those who perform the labor in the fields, because, to date, there is a dearth of food writing that approaches their labor conditions with the same research and sensitivity as that accorded to the health and well-being of livestock and the environment, broadly defined.[20] This is not surprising if we consider that the American public is not primed to evince concern for low-wage workers, especially immigrants. Moreover, there are good reasons why consumers prefer to imagine that a local farm is a shipshape family unit, with the exception of the harvest season, when from time immemorial some extra help is required. Correcting the record, as this book seeks to do, will help to raise the new food ethic to a higher level of integrity.

· · ·

The book begins with an examination of Hudson Valley food culture and an analysis of the legacy of agrarian ideology. In a brief history of the farm work force since the early 1800s, I argue that the region's small producers have consistently changed their farming practices to adapt to new challenges, including the almost-constant struggle to secure a stable and controllable labor force. In addition, the insecurity faced by Hudson Valley farmhands from the era of tenant farming up to today can be perceived as a form of subsidy to the farming industry. Moreover, I contend that the agrarian ideal led to farmers' common expectation that a cheap and viable agricultural labor market should be made available. By explaining how history and ideology influence the way that we think about farm labor today and offering a current snapshot of farming in the region, this chapter sets the backdrop for describing my field research.

Chapter 2 describes the living and working conditions of today's Hudson Valley farmworkers and examines the historical and structural factors inhibiting their ability to change their situations. I focus on the combination of their exclusion from labor laws, rural isolation, union neglect, and specific details about the status of the guest workers and undocumented immigrants.

The heart of this chapter explores how paternalistic labor control continues the long record of social mistreatment at the hands of the employer and the state. Finally, I recount the immigrant narrative of the newer Latino farm laborers with a focus on the contemporary factors that exacerbate their weak position. I conclude that for these typically undocumented workers, the realities of limited employability and anxiety about their legal status, combined with their aspirations to return permanently to their home countries—relatively new features of this agricultural labor market—make them largely accepting of the poor working and living conditions offered on the region's farms without much questioning. In particular, I show that workers' plans to return to their home countries allow them to rationalize their situations through habitual comparison with their lives there rather than with those of other U.S. workers.

Chapter 3 profiles Hudson Valley farmers as a struggling class of small producers who are chronically uncertain about the future of their enterprises, much like their historical counterparts. It presents farmers' perspectives on their primary challenges, such as the vagaries of weather, real estate development, corporate and foreign competition, low food prices, and difficulty in securing a stable, reliable work force. By personalizing the employers, I reveal the complexity of the labor landscape, paying particular attention to how agrarian ideology feeds into the management mentality of labor control. I argue that because farmers face multiple constraints over which they have little influence, they view the flexibility of labor costs as an important area of proprietary control. Consequently, they resist even minor proposed changes in farm labor policy for reasons of costs and to monopolize control of labor management decisions.

Rural New York was not a significant destination for Mexican and other Latin American immigrants until the late twentieth century. Chapter 4 explains this increased immigration and argues that farm owners have actively colluded with state bureaucrats in shaping recruitment to favor the employment of Latino workers and to discontinue the hiring of African American and Caribbean workers. These pages document the specific demand-side employer mechanisms through which Latino immigrants arrived in the Hudson Valley for low-wage work and analyze how employers' preferences for undocumented Latinos shape the migration stream in significant ways.

In addition, chapter 4 looks at how employers characterized these newer workers, particularly in contrast to previous groups of workers. My conclusions depart from those of scholars who explain the racial and ethnic characterizations of workers as rational employment decisions.[21] On the contrary, my findings support the idea that such characterizations have their own logic, related to race relations and the histories of immigration and low-wage employment in the United States and the Hudson Valley. That is why the quotidian experiences described by my interviewees proved so rich a source for explaining how racialization is reinforced among employers, their allies, and the public, further marginalizing the workers themselves.

Chapter 5 is a call for a comprehensive food ethic that works through legislative action. The alternative food movement is more visible than ever before, prompting my investigation for the possible role that food movement allies might play in promoting the rights of those who pick their food. It examines the record and challenges of farmworker advocacy in New York State, including the defeat of a pro-farmworker bill in the New York State Senate in August 2010, to reveal the obstacles to legislative change. This concluding chapter also outlines practical recommendations for how locavores might become more engaged in justice for farmworkers.

. . .

In her groundbreaking book *Agrarian Dreams,* Julie Guthman revealed not only that California's corporate agribusiness sector co-opted the organic standards of those who forged the pioneer alternatives, but also that the subsequent production and distribution processes of organics came to mirror some of the worst features of the global food system, including deceptive marketing practices, labor exploitation, and the creation of waste and hunger. She also revealed that these practices took place regardless of the scale of farming.[22] Guthman was not looking to discredit organic farms or to scare away consumers. Rather, her goal was to challenge the myths and misconceptions about the organic agriculture industry itself. *Labor and the Locavore,* which is inspired in part by Guthman's example and enthusiasm for food reforms, in no way seeks to diminish the consumption of local foods; I am an enthusiast of Hudson Valley farms and their produce myself. But by offering a much-needed analysis of the dynamics of the region's small-scale agriculture along with firsthand descriptions of the conditions of workers and the efforts of farmworker advocates to improve those conditions, my goal is to fully humanize the content and spirit of the reforms.

ONE

Agrarianism and Hudson Valley Agriculture

WHEN TODAY'S HUDSON VALLEY GROWERS are lionized in the pages of foodie magazines or the travel section of the *New York Times,* they are depicted as practicing a dying trade and preserving open space for the cultural and environmental good. Many of the region's farmers see themselves as part of a hardscrabble agricultural tradition (my own town in the region celebrates an annual "Hardscrabble Day"), and certainly their precarious economic position relative to owners of factory farms supports this perspective. Many of their ancestors came from very humble backgrounds, and some struggled against the oppressive tenant system of the eighteenth and nineteenth centuries. Though they may own hundreds of acres of land and hundreds of thousands of dollars' worth of farm equipment, their ability to stay afloat from year to year is never assured. Yet advocates of open space preservation see farmers' valiant fight to "hold on" as a defense against the developer's bulldozer.

At the same time, these farmers are enjoying the revival of interest in Hudson Valley agriculture by living up to the idealism of boutique farms, heritage fruit, pick-your-own venues, and branded products. Farmers' markets have proliferated all over the region, and numerous restaurants tout local products on their menus. "Violet Hill sunny side up egg with Berried Treasures ramps and Yuno Farms dandelions" was a featured item on the 2009 Earth Day menu at New York City's upscale Casa Mono. Restaurants such as Manhattan's Blue Hill—with a menu described as "seasonal American celebrating the bounty of the Hudson Valley"— design their offerings around the best local produce. Another ardent supporter of "hometown" ingredients, Rhinebeck's Gigi Trattoria in Dutchess County, boasts supplier relationships with forty local producers. These top-end promotions are evidence of a regional food culture in the making.

"Food culture" is presumed to be distinct from an "industrial food system." In this regard, the expectation is that the food be artisanal in nature and therefore associated with the labor economy of the expanded family, or the cooperative enterprise whose participant members hold some equity. Consumers of this culture are led to believe that their interaction with a small farmer is more akin to securing carrots and beets from a neighbor's flourishing garden than to the commercial exchange that takes place in a supermarket selling factory farm produce.

The regional marketing that promotes today's small farmers is replete with the echoes of more than two hundred years of agrarian idealism, a rich belief system that historically has sidelined the long-suffering but indispensable farmhands. Yet while it is common to hear today's farmers lament the difficulty of securing a consistent annual profit and finding suitable workers, the historical record shows there is nothing new about either of these complaints. Since their initiation into agricultural commodity markets in the early nineteenth century, the region's farmers have consistently transformed their growing practices in an effort to try to secure a profitable niche. Concomitantly, they have faced the challenge of securing a stable work force as changes in the nature of farming go hand in hand with changes in the labor market and growers' labor needs. Between the end of the era of self-reliance in the mid-1800s (when farmers depended on family labor or bartered community labor, and little monetary exchange took place) and World War II, the region's farmers, as a whole, did not enjoy anything resembling a stable agricultural economy or work force. This chapter shows how contemporary Hudson Valley food culture leans on the agrarian ideal and offers a snapshot of the history of farm laborers in the region, from the early decades of farmers entering into market systems to the current economic landscape of Hudson Valley agriculture.

HUDSON VALLEY FOOD CULTURE

There is no doubt that Hudson Valley farming qualifies as an exemplum of the artisanal, local growing ethos that has fueled the local food movement and new forms of agrarianism. Whether Hudson Valley food culture can be distilled into a brand-name dish, method, or product remains to be seen.[1]

Perhaps "local" is the best way to define the region's food culture, with its niche products, direct marketing, and accompanying promotional apparatus. "At its heart, a genuine food culture is an affinity between people and the land that feeds them," and it comprises "a set of rituals, recipes, ethics, and buying habits." This is Barbara Kingsolver's stab at defining the term and the ideal to which her family subscribed in their "year of food life."[2] A supermarket diet of multinational products and produce of unknown provenance is the very antithesis of what is imagined to be an authentic food culture, which involves knowledge about how foods are acquired, prepared, preserved, and consumed.

Such a sense of place is best represented by farm products, and Hudson Valley food culture is distinguished by its reliance on high-quality, fresh produce and meats, artisanal goods such as specialty cheeses, and boutique beverages like small-batch ales, hard ciders, wine, and other alcoholic drinks. One hallmark of local food here is heirloom and heritage fruits and vegetables, which have made a comeback in recent years, particularly since more discerning urban consumers have tired of commercially popular produce such as the Red Delicious apple, which has a perfect shape, robust color, and sweet taste, but whose overproduction has diminished its flavor. One of the region's current strengths is niche marketing of specialty fruit. A selection of these is offered at locations like Adams Fairacre Farms grocers (with three locations in the Hudson Valley), one of the first local grocery chains to aggressively promote regional produce. As autumn rolls around, shoppers will find clearly marked bins with detailed descriptions highlighting the qualities of more than twenty varieties of local apples. Bordering the produce aisles are jars of local foodstuffs: honey, jams, chutneys, salsas, marinades, and pickles. On the other side are arranged local milk, eggs, yogurt, and cider, while regional cheeses are to be found alongside gourmet selections in another part of the store. Although only a small percentage of the store's overall stock originates nearby, the range of products is a showcase of Hudson Valley agriculture—fulfilling the traditional function of the county fair—and the proportion of regional products is increasing annually. Adams grew out of an early-twentieth-century farm stand and exemplifies the type of regional economy that has allowed some of the valley's growers to thrive in the face of cheaper international products. In recent years large chain supermarkets, including Walmart, have begun to feature local produce in an effort to compete.[3]

The niche marketing of regional produce has been updated to accommodate new institutions and relations of consumption. Perhaps the most significant change in Hudson Valley agriculture at the end of the twentieth century was the development of opportunities for small producers to sell directly to consumers—a necessary strategy for promoting local food. The most visible institution, of course, is the farmers' market. Its progenitor, the Union Square Greenmarket in downtown Manhattan, was established in 1976 and quickly became the largest retail market in the state, featuring the stalls of as many as 140 regional producers a week. In the past decade the number of farmers' markets in the state has more than doubled, increasing from 235 in 2000 to 521 in 2012, with New York City's five boroughs accounting for 138 of these.[4] One of the newer models of local farming is the CSA (community supported agriculture), an alternative farming economy in which individuals receive farm products in return for an advanced investment or ongoing subscription, thereby providing the farmer with upfront funds needed to run the operation. With CSAs, individual members are small shareholders shouldering a risk in return for a portion of the yield. Not only have new farms established CSAs from the outset, but some longtime farmers have also added CSA subscriptions alongside their traditional retail and wholesale practices.

The culture of parochial purchasing extends well beyond the farmers' market since an increasing number of restaurants, as documented by their menus, have determined it to be a market necessity to offer local products. The region's food identity is anchored by the Culinary Institute of America (CIA), a prestigious culinary college in lower Dutchess County that has trained many of the area's chefs. Its courses include Food, Wine and (Agri) culture; Leadership and Ethics; and Ecology of Food. One graduate, Waldy Malouf, former chef at Manhattan's Rainbow Room, which he transformed during the 1990s, has been an important promoter of Hudson Valley food. The *New York Times* included Malouf's 1995 *The Hudson River Valley Cookbook* in its roundup of top cookbooks of the year, and it was also nominated for the prestigious Julia Child Cookbook Award, further institutionalizing public awareness of the region's offerings. An early innovator of American nouvelle cuisine was John Novi. His Depuy Canal House opened in High Falls, Ulster County, in 1969, featuring local food and earned a four-star rating in 1970 from Craig Claiborne, the *New York Times'* erstwhile dean of restaurant criticism. More recently, Laura Pensiero, owner of the aforementioned Gigi Trattoria in Rhinebeck, published *Hudson Valley Mediterranean* (2009), a paean to the region's local foods.

Prestigious restaurants are part of the promotional apparatus that helped give birth to a regional food culture in the Hudson Valley. Not only did chefs showcase local farms and regional cuisine, but they also contributed ideas and techniques to the common repository of local food culture. One chef in particular who has shaped ideas about the region's local food is Dan Barber, chef and co-owner of Manhattan's Blue Hill restaurant and proprietor of Blue Hill at Stone Barns at Westchester County's Stone Barns Center for Food and Agriculture, which includes a working farm, restaurant, and education center. Like Alice Waters's Chez Panisse Foundation, Stone Barns Center hosts a speaker series, conferences, a farmers' market, a curriculum for school gardens, and a plethora of programs for kids, such as summer day camp, story hours, cooking classes, and hands-on farm experiences. Other nonprofits that advocate for local agriculture include Putnam County's Glynwood Institute for Sustainable Food and Farming, which also operates a farm and fosters dialogue through events, publications, and local programs to promote sustainable farms and rural communities. The *Valley Table,* a Hudson Valley quarterly magazine dedicated to food and farms, is an important publication for taking the pulse of the region's food movement. In addition to publishing produce and restaurant profiles, some of them by local food activists and politicians, the magazine includes articles that promote the available range of farm products, from award-winning wines to artisanal cheeses and alcohol distilled from local fruit. In 2009, a new glossy, *Edible Hudson Valley,* began quarterly publication as part of a national network that now boasts eighty regional "community-based, local-foods publications" under the Edible Communities rubric.[5]

Finally, something must be said for the history and landscape that have shaped the region and legitimized its rural character. World-famous for the achievements of Hudson River School painters such as Thomas Cole and Frederic Edwin Church, it was one of the first American regions to establish tourism by taking advantage of the combination of natural beauty, relative proximity to New York City, and ease of transportation.[6] The combination of lush scenery, military history (West Point, home to the U.S. Military Academy on the west bank of the Hudson, for instance, was occupied by the Continental Army in the late eighteenth century), and arts contributed to the establishment of a high-profile cultural identity for the region, which included bona fide historical sites such as Native American settlements, Revolutionary War battlegrounds, and aristocratic riverside estates.[7] Employing the architectural and design talents of Alexander Jackson Davis and Andrew Jackson Downing, the owners of such estates, located mostly along the "aristocratic" east bank of the river, helped create the pastoral rural vernacular that defines the region as a whole.[8] Today, it is agritourism that draws visitors to the Hudson Valley; the food movement campaigns at the turn of the twenty-first century delivered a windfall of publicity and investment in the region's agricultural sector.

A variety of definitions of "local food" are in circulation. Some refer solely to geographic proximity (as in the one-hundred-mile diet), others to a type of distribution network like the farmers' market.[9] To the degree to which "local food" implies a food culture, this includes not only fresh, seasonal, diverse products, but also the active promotion of biodiversity and local sustainable economies. In and of themselves, these are important and valid reasons for why consumers might prefer to eat local. But characterizations of "local foods" do not stop there. For example, the Whole Foods website describes how local foods "enhance farming towns and their regional identities" Small local farms are a valuable component of a community's character, helping maintain agricultural heritage, preserve land use diversity and moderate development."[10] A more florid commentator on the Eat Local Challenge blog took it further yet: "Buying locally grown food is fodder for a wonderful story. Whether it's the farmer who brings local apples to market or the baker who makes local bread, knowing part of the story about your food is such a powerful part of enjoying a meal."[11] These more abstract associations—the stories told, the pride in a sense of place, and the high estimate of community and heritage—are components that feed into the heady romance of local food production.

Underpinning the romance is an automatic equation of geographic proximity with goodness, a phenomenon termed the "local trap" by scholars who argue that scale—whether local or global—is always socially constructed and nonuniform.[12] The conflation of localness and wholesomeness is strongly echoed in the writing of food movement leaders like environmentalist Bill McKibben, novelist Barbara Kingsolver, and ethnobiologist and nutrition ecologist Gary Nabhan, all of whom have drawn up manifestos from their locavore, neoagrarian adventures to reinforce these beliefs. In addition, a generation of scholars have expounded on the positive aspects of local food systems, which include economic and social benefits,[13] the promotion of justice and community through face-to-face interactions with food producers,[14] and the capacity of alternative agrisystems to promote civic engagement and enhance democracy.[15] The most recent lineage for this revival dates to the "natural foods" revolution of the 1960s, fostered in communities in Vermont and Northern California, institutionalized in food co-ops, and rooted in

E. F. Schumacher's "small is beautiful" philosophy of human-scale technologies.[16]

THE AGRARIAN IDEAL

Although this book is primarily a contemporary survey of agriculture and agricultural labor issues in the Hudson Valley, I want to emphasize how much history and ideology have shaped the way that we think about these topics. The contemporary subjects of my research have their own individual stories, which are explored in the chapters that follow, but the constraints they face resonate with powerful legacies from the record of agricultural history.

It would be difficult to overstate the importance of agrarianism not only as a formative component of national ideology, but also as a determinant in the political economy of food. Whether consciously or not, it has been adopted by all types of farmers (organic, local, very small, corporate) and is deeply embedded in the public mind.[17] The values attached to American agrarian life are associated with high moral virtue, economic self-sufficiency, and individual freedom.[18] U.S. agrarianism posits small-scale family farming as the basis for a model society, as articulated in Jefferson's influential vision of a nation of freeholders occupying the middle landscape between cities and wilderness. Jefferson's model was fueled by the ideal that "those who labor in the earth are the chosen people of God."[19] Other canonical writings, such as gentleman farmer J. Hector St. John de Crèvecoeur's *Letter from an American Farmer,* helped establish agrarian self-reliance as a distinctively American trait.[20] (Crèvecoeur originally farmed in the Hudson Valley before moving south to the Carolinas.)

The agrarian ideal, also referred to as agrarianism, romantic agrarianism, and the agrarian myth, encompasses three main beliefs: farmers are economically independent and self-sufficient; farming is intrinsically a natural and moral activity; and farming is the fundamental industry of society.[21] These three tenets feed into each other. For example, discussions about farmers' self-reliance are often seeded with ideas about their being virtuous even though there is no inherent connection between the two. In addition, the belief in the wholesomeness, reliability, honesty, and hard manual work of farmers, as well as the role they play in defending agricultural traditions, feed into the nobility bestowed on their self-reliance. This self-directed toil is not only basic to society at large, but it is also perceived to be the lifeblood of a republican democracy. Indeed, the freehold farmer has historically been cast as the ideal citizen who has embraced his moral duty to society and the common good.

The moral ideal of agrarianism in the United States was largely spread by farm leaders in agricultural journals and other outlets.[22] Agrarianism offered what Richard Hofstadter called a "standard vocabulary to rural editors and politicians" so that it could be as easily preached on campaign trails as in local newspapers.[23] The virtues of farming were commonly touted by both government and private farm organizations. Despite the fact that agrarianism rests upon the model of the subsistence family farm, as opposed to the modern, commercial farms that developed in the late nineteenth and early twentieth centuries, it became the prevailing ideology of agriculture as a whole, even as highly mechanized agribusiness moved further and further away from the original ideal.[24] Today, perhaps only the ideal farmer proposed by Wendell Berry fits with the foundational ethos of agrarianism—a subsistence grower who does not rely on external labor and is more interested in sustainability than profit. Agrarianism allowed agriculture to hold a privileged place in U.S. society at large, but more importantly it continues to guarantee government attention to those whose livelihood and profits are bound up in cultivating the soil.[25]

When the nineteenth century ushered in new markets and agricultural practices that rapidly shifted the focus from subsistence to profit, the self-sufficient farmer who relied on community mutual aid became a businessman and a land speculator.[26] Farmers struggled to uphold the practical prerequisites of the agrarian codes,[27] but the ideology proved very serviceable to the industry as a whole, particularly in the name of agricultural exceptionalism: farming was a "unique" industry because it fed the nation.[28] The utilitarian value of agrarianism survived the twentieth century and harmonized well with the energy of the environmental movement. From the gentleman farmer on his exurban acres[29] to the urban farmers of downtown Detroit,[30] the agrarian fantasy is alive and well.

Agrarianism has served various and often contradictory ends. The ideal has even pitted farmers of different types against each other by offering a flexible rhetoric to anyone laying claim to it. In this way, it has served landowners both large and small, subsistence farmers, antisprawl environmentalists, utopian communalists, and even farmworkers themselves. For example, while some would-be agrarians have promoted progress and efficiency in order to defend commercial farming, others have focused on the use of agriculture to serve environmental and communitarian goals.[31] Although it would seem that those with a direct line of descent from the yeoman farmer would hold the most legitimate claim to agrarianism, a claim foregrounded by the antiforeclosure Farm Aid campaigns of the 1980s, those who have benefited most from the ideology are the largest, profit-centered agricultural producers.[32]

Agricultural exceptionalism has driven public attitudes and policy decisions about farming for more than two hundred years.[33] Historian David Danbom argues that one reason agrarianism is so compelling is that it helps to recapture lost innocence.[34] The public elaboration of agrarian beliefs has been prolific and pervasive; agricultural economist Don Paarlberg has claimed that these beliefs don't need to be learned by those growing up in the country, but instead they are merely absorbed.[35] This exact experience is explained by conservationist and food writer Wendell Berry, who describes being an agrarian from an early age, although he did not learn the term until he was in college.[36] Agrarianism still resonates today, and food writers trade on the romance of agrarian ideals.[37] Their florid descriptions of farmers' markets, glowing profiles of local purveyors, and anecdotes of happy farm animals put a gloss on the public image of the food movement.

Farmers have called upon the singularity of agriculture as well as the cultural cachet of agrarianism to influence policy in favor of their interests.[38] It is widely understood that agriculture merits a range of support from government because it provides vital sustenance to the U.S. population. No other industry, farmers and their organizations often contend, faces the same demands or deserves similar treatment, particularly in regard to safeguarding its cheap labor supply. A belief in agricultural exceptionalism helped to cement the agreement that excluded farmworkers from New Deal labor protections, such as the collective bargaining protections of the National Labor Relations Act (1935) and the minimum wage and overtime laws of the Fair Labor Standards Act (1938).

Unlike in other industries, and due to government support for the agriculture industry, the increased costs of doing business in farming are not all passed on directly to the consumer. Most farms lose money, and the U.S. government compensates for these losses through heavy subsidies, providing between one-quarter and one-half of farms' net income. Large and corporate farms benefit the most from agricultural subsidies, particularly under Farm Bill and emergency relief provisions. Farms that receive the least in government payments are those growing fruits and vegetables.[39] Government payments, however, are only one piece of the subsidy process. There is an intrinsic relationship between the success of this regional food culture and the exploitation of farm laborers; arguably the largest agricultural subsidy in the region comes not from government, but from farmworkers themselves. According to a 1994 U.S. Department of Labor Report, it is farmworkers who subsidize farms "with their own and their families' indigence," through structures that "transfer costs to workers" and leave them impoverished.[40] The cumulative impact of the wage discount—maintained so low by government policy—is vital for keeping the industry afloat.[41]

As the core signifier of agrarianism today, the "family farm" evokes a household unit of production with perhaps a few hands to assist at harvest time or perform odd jobs. Technically speaking, however, the term is much more elastic in definition. The designation accepted by Congress excludes only nonfamily corporations, although another congressional definition limits family farms to those without a hired manager and with fewer than 1.5 hired workers annually.[42] Alternatively, the USDA Economic Research Service bases its classification on ownership and management structure. It defines family farms as those legally held by a single family with control over the way farm profits are spent. At the same time, it excludes nonfamily corporations or cooperatives, as well as estates, trusts, and farms run by hired managers. Under this rubric, 98 percent of U.S. agricultural operations are categorized as family farms.[43] According to census data, 84 percent of New York's farms are individually or family owned, but in its promotional material, the New York Farm Bureau asserts that 99 percent of the state's farms are "family owned" and refers frequently to its "member families." This statistic likely includes all farms—including partnerships, family-held corporations, and trusts—except for nonfamily-held corporations, following the lead of Congress.[44]

Using definitions based on ownership and management as opposed to income, number of workers hired, or inclusion of subsidiary businesses means that New York's largest farms can plausibly be marketed—both by the growers and by agricultural organizations—as family farms. For example, Torrey Farm, one of the state's largest at more than ten thousand acres, employs several hundred workers, and, according to Dun & Bradstreet, its annual sales exceed $21 million.[45] Because it is owned by one family, however, it is as much a New York family farm as a twenty-acre arugula plot with no hired hands. More to the point, the "struggling family farmer" defense that is typically mounted against labor reforms masks the reality of the agricultural labor market in New York since only one-quarter of New York's farms actually hire workers.[46] In other words, three-quarters of the state's farms rely only on household labor, but they are categorized alongside Torrey Farm as family establishments. This is one of the ways that agrarianism in name only facilitates profits for the largest agribusinesses in the state and throughout the country. The romance will not wither away as long as it proves serviceable to an industry that profits from the venerated position of agriculture in our society to the detriment of small-scale farms.

While agrarianism might have its most potent symbol in the family farm, the organic movement was also fed by its ideals.[47] The communitarian and owner-operated nature of early organic farms fit neatly with the wholesomeness and farmer independence promoted by agrarianism. Guthman argues that the new agrarianism was anchored by the organic movement, which combined environmental concerns, social justice, and family-scale farming, although, as she herself points out, the social justice concerns did not include attention to hired hands.[48] Guthman debunks the notion that organic farmers operate according to motivations different from those of conventional farmers. It is commonly assumed that their operations are distinct from the industrial food system or that organic farms serve the social good while conventional farmers are only interested in profit.[49] Organic produce is marketed successfully as a value-added product, and such a process relies to some extent on scarcity, for if too many other products have that same quality, the valorization of organic produce is not successful. Yet the USDA organic standards were co-opted by industrial agriculture to ensure they fit with the needs of big agriculture, which could then capture a market share. In response, diehard organic farmers called for a new strategy to promote their farm products, arguing that scale was a vital component; according to their logic, the "small-scale family farm" became "a proxy for social justice."[50] To some extent this strategy promoted a shift in attention from organics to local food, since the latter's direct-to-consumer marketing model is much more difficult for industrial agriculture to adopt.

Agrarianism also helped perpetuate the notion that labor was treated better on organic farms, if for no other reason than that labor concerns were initially not addressed by food writers or scholars. Other commentators have shown how images of pastoral beauty hid the migrant laborers who helped to create it and who live on "the other side" of the sculpted landscapes.[51] Indeed, the perception that the agrarian realm was a modern-day Eden relied entirely on the invisibility of migrant workers. Whatever attention was placed upon workers offered them up as "noble savages" who had modest needs and did not mind dwelling on the rural periphery in humble living conditions.[52]

New York's Hudson Valley in particular is home to exactly the sort of farms upon which the nation's early republican ideals were founded, giving growers and industry lobbyists all the more reason to use the mystique of agrarianism to serve their interests in the public sphere and at the state legislature. Consider the following quote extracted from the "Grassroots" newsletter of the New York Farm Bureau, the foremost interest group serving farmers in the state: "The most important thing that comes from our farms is the quality of citizenship that grows there. I see it in the responsible, 'can do' attitude in these kids. So many of them seem mature beyond their years. . . . Farming imposes a code of conduct on a person. It is called responsibility. It is called self-reliance. . . . This reservoir of responsible citizenship is, to me, as precious and valuable as the land, itself. Maybe more so."[53] Rhetoric like this is carefully crafted not only to nurture public perceptions about agrarianism, but also to win a place in the speeches of the lawmakers who are targeted by the New York Farm Bureau. As for farmworkers, the organization highlights labor as a top policy priority, while success in securing workers and promoting a legal work force is cast as a survive-or-die option. According to the organization's magazine, laws such as overtime protections for farmworkers would cause farms to "become something less bucolic." The reader is supposed to make the connection that unless farmers' labor priorities are attended to, the state will lose its picturesque farms and all that accompanies them. In other words, legislation friendly to farmworkers would destroy farms.

Today the Hudson Valley signifies a respite from the ills of corporate agriculture in much the same way that California in the nineteenth century offered a potent dream of pastoralism that drew waves of opportunity-seeking immigrants. Both regions depend on agrarianism to promote and market their farm products. Today it is local food that gets linked to "community capitalism," a kinder, gentler approach to business that incorporates civic engagement.[54] Kingsolver goes even further in suggesting that "'locally grown' is a denomination whose meaning is incorruptible."[55]

HUDSON VALLEY FARMWORKER HISTORY

Browsing through the regional section of any Hudson Valley town library shows that the histories of New York and Hudson Valley farming have received substantial attention by a range of writers, from popular commenta-tors to researchers and policy analysts. Within these volumes, however, com-paratively little attention has been given to the agricultural work force, except to record data such as wages, labor shortages, or the volume of workers. In part this is due to the dearth of archival material.[56] Farmers leave records of their activities, while hired field hands do not. Moreover, research conducted on the state's farmworkers is dominated by labor market analysis and is often disconnected from the larger context of agricultural trends.[57] Important recent studies address the implications of the ethnic shift from black to Latino farmworkers in the late twentieth century, focusing on the integra-tion of new immigrants into local communities and farm management prac-tices tailored specifically to them. However, these studies fail to discuss the shift in its fuller historical context.[58] The history of the relationship between the development of Hudson Valley agriculture and its work force is impor-tant for understanding the racialization of farmworkers and how they devel-oped as a distinct class from farmers and why farmers expect a ready and cheap supply of workers. History also explains that the region's farmers have experienced significant instability in their ability to secure workers and to make a profit, and for almost two hundred years they have creatively responded to varied constraints by changing their practices.

Hudson Valley agriculture, along with Northeast agriculture on the whole, developed along lines quite distinct from those in other parts of the country. Urban centralization and early industrialization in the Hudson Valley constrained the growth of large-scale industrial farming. In contrast, the southern colonies, Northwest Territories, lands of the Louisiana Purchase, and even the irrigated West all offered better conditions for settle-ment on huge tracts of land. In addition, Hudson Valley farmers faced rocky and hilly terrain, a short growing season, and unpredictable weather. Later, intense competition, the cost of hiring labor, and the pressures of suburban-ization would further challenge the region's growers. With its great estates, the Hudson Valley was dominated by a manorial economy through the mid-nineteenth century; a handful of families owned expansive areas of land and rented small landholdings to farmers.[59] As a result, a neofeudal landholding system developed in contradistinction to the industrial plantation system that pervaded the Eastern Seaboard of the Americas from northern Brazil to the Chesapeake.

The small size of farms and meager existence of leaseholders meant that few non–family members provided labor. For extra labor during the harvest, farmers usually turned to neighbors, especially since scarce contract workers were perceived to be too expensive, difficult to discipline, and morally untrustworthy.[60] Nonmonetized neighborly cooperation, which was carefully tracked,[61] fostered local solidarity, fueled pride in self-reliance, and boosted the practices and ideology of community mutual aid.[62] Thus was the communitarian side of the agrarian ethic lived and affirmed. Local artisans, who did not have farms or who had very small holdings, also relied on social connections to supplement their work with day-labor agricultural pursuits.[63]

The 1820s saw the region's farmers drawn into agricultural markets as they gradually turned to hired labor to accommodate the transition to cash crops.[64] Early industrialization brought textile and other factories to the region, which stimulated agricultural production in response to an increasing volume of industrial workers who required a local food supply.[65] Although factory production reshaped the landscape by creating urban centers populated by the new working class (Fishkill, Poughkeepsie, Hudson, Kingston, Newburgh), the anchor of the Hudson Valley economy was still agriculture, in contrast to the ascendancy of manufacturing in the New England states.[66] When the Erie Canal, opened in 1825, it brought fierce competition from Midwest wheat farms, and the region's farmers shifted their primary production from wheat and meat to dairy, sheep (for wool), beef, and poultry. By the mid-nineteenth century, Hudson Valley farms were also cultivating orchards, largely for cider. One writer described the "bewildering variety of individual and local adjustments" made by the region's farmers in the period between 1790 and 1860.[67]

One backdrop to this volatile but resilient era was the persistence of the manor system. The Anti-Rent War of the 1830s and '40s, involving ten thousand tenant families at its peak in 1845, led to the dismantling of the manorial land economy.[68] Freeholders who earned their way out of the bonds of tenancy tended to stay on their small tracts of land, thus preserving the pattern of small holdings in the region. German and Dutch surnames from this period can be found in the roll call of today's farmers. More potent yet is the persisting survivalist legacy of the hardscrabble profile embodied by the iconography of the small farmer under threat of foreclosure.

Self-reliance was reinforced by the necessity of maintaining local ties in communities with few outsiders. In 1845, for example, 93 percent of Dutchess County residents were locally born,[69] but that was before the railroad reached Poughkeepsie from New York City at the end of the decade and began to change the demographic profile of small rural towns. The trains facilitated the arrival of Irish, English, Scottish, and German immigrants as well as African Americans in search of farm work.[70] Prejudice against blacks and stereotypes about the European immigrants generated community hostility and served to cast farmhands as a different class from farmers.[71] More significantly, the farmworker as a job category was racialized.

Hired hands were much more common by the mid-nineteenth century, although farmers were still very sensitive to labor costs. To avoid wage inflation through competition, some growers established a common pay rate for laborers who were not family members.[72] Workers were also offered to neighbors for a fee, a sign of waning community reciprocity.[73] Local youth were a significant source of labor and often served as very low-paid apprentices, occasionally receiving only room and board, as many hoped to start their own farms by climbing the agricultural ladder. Census data for several Hudson Valley towns from 1850 show that 20 to 37 percent of farmers hired unrelated workers.[74] According to the same records, one-third to one-half of midcentury families in Dutchess County were landless farmworkers.[75]

By midcentury the Industrial Revolution was in full swing, radically changing the nature of work and employment. Kingston, on the west side of the Hudson River, was the center of Fulton's steamboat production, and Poughkeepsie, on the east side, had mills in full operation as an important manufacturing center. In the eighteenth century, all but 10 percent of Dutchess County residents were engaged in agriculture, but by 1850, one-quarter of the county's residents worked outside of farming.[76] The resulting changes in agricultural markets and production altered farmers' needs for, and relationship to, workers. Farms became more efficient with the introduction of new scientific methods such as crop rotation and commercial fertilizers, in addition to the use of modern equipment like cultivators, seeders, threshers, and reapers.[77] As a result, fewer workers were required outside harvest times, a situation that persists today due to the fact that fruits and vegetables are largely handpicked. The increasing demand for seasonal harvest labor altered the job category of farmworkers since it would become increasingly difficult for so many harvest farmhands to piece together year-round work. At the same time, low-skill jobs on the railroad and in construction lured workers away from agriculture with the promise of higher wages.

New farming technology and improved transportation—barges and railroad—made the mid-nineteenth century a golden age for Hudson Valley growers, but with the completion of the first transcontinental railroad in 1869, agriculture was once again transformed. Wheat from the Midwest bonanza farms flooded the Northeast market and vanquished the region's grain commodities once and for all. Hudson Valley farmers had to turn to fruit and dairy. Notwithstanding the perceived financial and social costs of hiring workers, labor needs increased since cultivating fruit was labor-intensive and animals required year-round attention. By contrast, wheat had been a low-maintenance crop. Even so, the seasonal nature of fruit and vegetable production required a good deal of labor flexibility. With few, and largely seasonal, positions to offer, farmers lamented the end of community reliance, and with little market experience, they struggled to make ends meet. Any outlay of cash was a risk that threatened to push the farm into the red.

Although farm labor wages peaked during the Civil War, the ability to demand high wages due to a reduced labor market was short-lived, and wages did not return to that level until the early 1900s.[78] Following the war local labor returned, and the volume of immigrants hoping to settle in the valley rose. One local author mentioned the surplus of farm boys seeking apprenticeships in the 1880s, likely for room, board, and very low pay, and at the same time he repeated the common refrain of how farm wages had been driven up by better-paid local urban employment.[79] By the turn of the century, excess labor, probably due to the severe depressions of 1870s and 1890s, had reduced wages in every sector. The apprenticeship experience fed the ambitions of youth who hoped to establish their own small farms. Yet by the late nineteenth century the shrinking volume of available land for sale restricted this option, particularly in the lower Hudson Valley due to the pressure of early suburban development.[80] Employers fueled their own agrarian dreams by hiring cheap labor, while the farmworkers rationalized their low-wage "training" in the name of their futures, a dream of upward mobility that is still very much alive among immigrant farmworkers today.

By the turn of the twentieth century, the agricultural ladder was broken. Farmhands lost the opportunities available to previous generations, and seasonal work became normative and permanent.[81] Economic and class disparities between farmers and their laborers widened. Earlier generations of workers had been housed in farmers' homes, married farmers' daughters, and had become fixtures of small rural communities. That changed with the advent of the businessmen-farmers who, no longer sympathetic to the plight of farm-hands, began to identify workers' interests as hostile to their own and associated workers with both discipline problems and high costs.[82] Farm journals reinforced the new employer attitude as they encouraged managerialism and frequently addressed the "farm labor problem" in their pages. The agrarian legacy helped to naturalize the stratification, differentiating between the local and foreign work force, to the detriment of the latter.[83]

The end of the nineteenth century saw the development of fruit and vegetable truck farming and the establishment of labor camps to house seasonal workers. While rail lines and waterways offered access to many markets, trucks opened up new towns for delivery, and farmers did not have to accompany their goods, as the truckers acted as middlemen. Hudson Valley producers were diversifying their crops, shifting to orchards, berries, and vegetables, and some turned to specialty crops; grapes, for example, would become the largest Hudson Valley crop through the 1920s. Farmers individually negotiated to sell their fruits and vegetables to trucker-wholesalers for resale in urban markets. Consequently, competition between farm families resulted in a further diminution of community mutual aid.[84] Truck farming increased the need for seasonal workers, which waves of European immigrants would provide, and Ellis Island is, after all, but downriver from the heart of the Hudson Valley. Farmers hired first-generation immigrant workers, including many Italians, from nearby urban centers, a pattern that lasted for several decades. Italians found a niche as labor contractors, some later establishing their own farms.[85] In addition, unemployed mine workers from Pennsylvania were an early migrant stream of interstate workers to come to New York. They were joined by traditional transients, or "hoboes," who followed the rail lines that were now well established.

When migrant workers first began to work on New York farms, it was common for them to find housing with growers and community members, a pattern of the mid-nineteenth century that changed as labor camps developed. The state's first migrant camp was probably built between 1910 and 1915 for central New York bean and pea pickers.[86] These labor camps altered the relationship between workers and farmers by sealing their spatial polarization. As in an inner-city ghetto, farmworkers were relegated to a segregated location, further distancing farmhands from local communities and providing them a different standard of living than their employers. More than anything, the labor camp firmly demarcated the farmworker job category from the status held by employers and other locals.

During the early twentieth century, the economy of truck farming and a continuing labor surplus allowed for the relative success of many Hudson Valley farms. Increased immigration to New York City meant increased demand for food. From 1910 until 1919, farmers in New York received good prices for their produce, particularly as World War I raised the price of farm products.[87] Local fruit growers had a competitive advantage for some years before refrigerated railcars (standardized around 1905) allowed for the transportation of sensitive fruits from larger out-of-state farms. Outside metropolitan areas, truck routes remained the favored link to markets for several decades as refrigerated tractor trailers did not prevail in all regions until mid-century. With technological advances however, California fruit began to arrive on the New York City market with only a twelve-day transit, and Hudson Valley growers again had to rethink their business plans and adapt to a transformed market.

New challenges from Western competition were compounded by the flight of local workers and family members to urban locations. Moreover, with the onset of the First World War, able-bodied farm men left as soldiers, European immigration was curtailed, and migration from the South was very limited. In response, the government began to play a role as labor recruiter. Federal programs helped supply foreign workers to New York's farms, mostly from the West Indies, but also from Canada. In addition, the U.S. Department of Labor organized one million male youth into the United States Boys' Working Reserve, who staffed mostly northeastern farms.[88] Aggressive recruitment of local labor also helped agriculture, including work-or-fight laws that were essentially labor drafts, enforcing work or jail for those not in the armed services.[89] In addition to taking advantage of government programs, farmers began actively to recruit southern workers.[90]

From 1920 to 1940, New York experienced an agricultural depression, causing the price of farmland to decline drastically. During World War I mechanization slowly expanded and farmers began to use tractors, thus reducing the demand for labor.[91] The use of southern migrants increased, as did hiring immigrant labor from urban centers. Southern blacks, both single males and families, arrived in the Great Migration of the 1920s. By the end of the 1920s, out-of-state migrant workers had become an integral component of New York agriculture.[92] The Great Depression further enhanced the farm labor supply as a result of urban unemployment, and, due to reduced work opportunities elsewhere, many locals stayed on to work the land.[93] I interviewed one farmer born in the mid-1920s who recalled how desperate many were for work during this period. The ferry across the river had its first crossing at 6 a.m., but that would put workers on the farm too late to secure the best berry rows for picking. To arrive early enough, he explained, "These guys and gals would come over in a rowboat to pick strawberries. . . . Sometimes there was more people than would fit in the boats they had, and so some would swim over and hang on the boats, even on a cold morning." Yet, when the worst of the Depression ended and hiring started up in the cities, it once again proved difficult to attract urban workers to the farms.

World War II was a turning point in Hudson Valley agriculture. The war stimulated the state's agriculture sector, and 1940–49 saw a new round of investment in agricultural technologies.[94] The government also began to ramp up its recruitment of farmworkers to meet the demand for supplementary labor. The Farm Manpower Service, established by a resolution of the New York State War Council in 1943, supplied 375,000 domestic and foreign seasonal laborers for New York's fruit and vegetable farms from 1943 through 1945.[95] Much of this recruitment aimed at helping the war effort by securing local food systems. The vast majority (89 percent) of the supplemental labor force was local, and most of the others were African Americans from the South. A great diversity of workers, including soldiers and sailors, conscientious objectors, and patients from mental health institutions, provided relief during World War II.[96] Other sources of labor included immigrants on vacations from the city who longed for the rural settings of their home countries.[97] Aside from local labor, foreign workers, predominantly from Jamaica and the Bahamas, Canada, and Mexico, worked on New York farms during the Second World War.[98] The war also prompted the formalization of programs for guest workers from the West Indies and Puerto Rico, programs that continue today. Additionally, the U.S. War Department supplied more than ten thousand German and Italian prisoners of war who worked farms while being housed in New York's prison camps.[99] Labor was not in short supply as long as the state acted as padrone and workers saw their efforts as a patriotic duty.[100]

The midcentury years saw another radical shift in agricultural practice. Even though the number of farms and amount of farmland declined significantly—largely due to suburbanization—new technologies generated higher crop yields than ever before. However, despite such advancements, the harvesting of fruits and vegetables still required manual application due to the sensitive nature of the goods, which were easily damaged. As a result, seasonal labor needs became much more concentrated during the harvest. Postwar workers were still mostly local, yet the volume of southern migrants grew by 140 percent between 1943 and 1948.[101] Also included in the postwar New York seasonal agricultural labor force were urban workers from Slavic and Mediterranean countries,[102] including Poland and Syria.[103] Pennsylvanian families still migrated to New York for work, and foreign guest workers included Jamaicans, Bahamians, Puerto Ricans, and Canadians. One account of farm labor in the postwar period records "interstate" migrants from Pennsylvania, Kentucky, West Virginia, Tennessee, Oklahoma, and Arkansas and distinguishes them from "Southern migrants."[104] This categorization leads one to presume that the interstate migrants were whites and the southern migrants were African Americans, considering that several of the states listed as supplying interstate migrants are indeed southern, but not, in fact, categorized as the "Deep South," which had a higher concentration of blacks than the "Upper South."[105] This is a distinction that reflects the racialization of "types" of workers that had increasingly become the rule of labor segmentation on the farm, in both housing and task allocation. African American migrants were often seen as ill suited for any other work, and single males were characterized as shiftless nomads, whereas the white workers, especially those with families, were understood to be economic refugees excluded from a reorganized southern agricultural system.[106]

The Northeast saw an earlier, more extreme loss of farmland than the rest of the country, in part because of competition from the sunbelt's output of fruits and vegetables. However, this was certainly not the only factor involved, as the development of subdivisions posed a new threat to agriculture. The postwar years in the Hudson Valley saw returning soldiers eschewing farm work for better jobs as they moved into newly developed suburbs.[107] Home buyers were drawn to the bucolic landscapes, but they were unaccustomed to the sounds and smells of farming and voiced their complaints accordingly. Agriculture in the Hudson Valley was now on the metropolitan fringe or the "suburban frontier," as one of the farmers I interviewed put it. This presented specific challenges as farmers attempted to accommodate (or antagonize) their new neighbors, creating zones of low-intensity conflict.

At the same time that a middle class of nonagricultural workers was becoming established, the Hudson Valley saw high rates of net unemployment from the 1950s to the 1970s as manufacturing moved south. The farmworkers, however, were increasingly from out of state. From the post–World War II period though the 1970s, African American migrants, predominantly from the South, were the largest group of migrant farmworkers in New York. The pattern of southern migrants' annual return to New York was established through their participation in the eastern migrant stream.[108] The high concentration of African Americans in agriculture on the East Coast mirrored the racialization of farm work that began in the South in the pre–Civil War era. Child and youth labor were also important sources of farm labor, as workers traveling in families worked together. At midcentury, the West Indian guest worker program was extended and focused on Jamaica, providing workers along the Eastern Seaboard, including for the apple harvest in New York.

By 1960, New York was employing 27,600 interstate farmworkers who were almost exclusively African American migrants, including youth.[109] Puerto Rican workers also staffed these farms.[110] In the course of the next decade, however, mechanization displaced many migrant workers, and the number of New York's seasonal workers dropped by half.[111] The volume of African American migrants further decreased due to urban migration and increased job opportunities in southern states, particularly in the service and construction sectors in Florida. Technology that extended the orange growing season further reduced the migrant stream traveling from Florida. Moreover, the children of southern workers were receiving a better education than previous generations, and as a result few followed their parents into farm work. During this same period, Caribbean workers who first arrived as guest workers began to settle in the United States. They would have been undocumented when they initially broke their contracts, and many married U.S. citizens and gained their green cards. They traveled the same migrant stream as African American workers. The 1970s saw a steady decline in the volume of Puerto Rican farmworkers, and relatively few are found today in New York, in contrast to the concentration of Puerto Rican farmworkers in New Jersey and Massachusetts (a group that deserves further study). Anecdotal evidence also suggests that local teenagers, a previously reliable source of local seasonal labor, began to opt out of farm work toward the end of the twentieth century.

After four decades of New York farms hiring predominantly black workers—African Americans, Caribbean workers who settled in the United States, and Caribbean guest workers—for seasonal labor, a shift toward Latino workers gained momentum. With fewer African American and Puerto Rican workers being hired in the 1970s, New York farmers increasingly turned to Jamaican workers, both guest workers and those settled in the United States. Haitians fleeing Jean-Claude Duvalier's oppressive policies also had a presence on New York farms in the 1980s, and there are still some Haitians working on farms there today. By the late 1980s, farmers began to rely on a newer population stream: Latino workers. Three decades prior, small numbers of Mexican and Chicano workers, most of whom were probably American citizens, traveled to New York from other parts of the United

States for farm work, settling in New York City and the lower Hudson Valley.[112] Latin American immigrants became a more significant part of the New York farm work force in the late 1970s, and their numbers have risen steadily ever since. These Latino newcomers were arriving undocumented and, increasingly, directly from Mexico and Central America.

This brief history highlights the instability of Hudson Valley agriculture. Ever since the early nineteenth century, farmers have almost constantly needed to adjust their operational practices to respond to market changes and new forms of competition. Although they have successfully restructured in order to survive and even flourish, they have been forced to do so repeatedly. An equally volatile pattern is evident in the long history of changes to labor needs and the regional labor market. Although structural factors have allowed farmers to hire cheap workers at times (during agricultural and economic depressions, when the government played a role in supplying workers, and after new mechanization), wages have increased at other times (during wartime and due to rural flight, the lure of nonagricultural jobs, and the increased demand for harvest-only labor). Records show chronic instability and turnover in the farm work force (family labor, community help, European immigrants, African Americans, apprentices, immigrants from urban centers, interstate migrants, guest workers, government-supplied workers, and Latinos). These factors have impacted the profitability of farming from one generation to the next. During periods in which labor costs seemed to cut into profits, employers developed resentment toward their hired hands. In addition, the period of community reliance was so formative in the development of Hudson Valley agriculture, and the use of apprentices who would work their way up to be farmers was common enough, that these led to an expectation of a work force that was cheap and readily available. Such expectations would be passed down from one generation to the next. Perhaps the most prominent characteristic of the region, however, is the racialization of workers in the nineteenth century, with their identity as a distinct class becoming institutionalized by the twentieth century.

THE CURRENT STATE OF
HUDSON VALLEY AGRICULTURE

In addition to the shifts in the labor market that I have described, the late twentieth century also saw substantial changes to agriculture in the Hudson

Valley. Although the 1980s brought small farms to their knees all across the country, land prices did not fall in New York because of the expanded interest in the land by real estate developers.[113] Higher land prices meant higher taxes, and the resultant economic stress gave farmers an extra incentive to sell to developers. Although the general twentieth-century decrease in New York farmland and cropland had largely peaked by 1969, the 1970s and 1980s showed a much greater loss of farmland and cropland in southeastern New York, including the greater Hudson Valley and Long Island, than in other parts of the state.[114] Developers' pressure on the lower Hudson Valley has been a longstanding trend due to its proximity to the city, and the dairy sector in particular has been greatly affected. One former dairy farmer put it succinctly: "from dairy to hay to horses or houses."[115] In general, the trend at the turn of the millennium was for land still in agriculture to be used for purposes more profitable than food production. These included nurseries and the cultivation of sod and flowers instead of fruits and vegetables.[116] In the 1970s Dutchess County had the largest number of dairy farms in the state with 275, yet by 2009 it had lost about 90 percent of these, the number reduced to 25.[117] Some of these holdings indeed became equestrian ventures and housing developments, while others became vegetable farms. Despite the consolidation and corporatization of its farms in the late twentieth century, small holdings still predominate in the region due to geography and development patterns. Forty percent of the region's farms have fewer than fifty acres.[118]

By 1997, the loss of farmland to developers spurred one farming organization to describe the Hudson Valley as the one of the most threatened agricultural regions in the country.[119] Antigrowth measures are one response to the surge in land development.[120] Such efforts provide much-needed funds to farmers by compensating them for eschewing the right to sell land for development, through the government-, private-, and nonprofit-funded Purchase of Agricultural Conservation Easements (PACE) and Purchase of Development Rights (PDR). Farmers receive a lump sum in return. In addition to development pressure and overseas competition, today's Hudson Valley farmer faces other significant challenges, including high property taxes, unstable commodity prices, the continued loss of young people to more lucrative occupations, and the ever more volatile climate.[121] Fluctuations in both temperature and rainfall, often attributed to climate change, have become more pronounced. Veteran farmers, with a keener memory recall of weather than most of us, characterize these extreme conditions as unprecedented.

"Agriculture is New York's number one industry." In the course of my research, I came across this phrase and others like it again and again in public relations material for the state's agricultural sector. Despite the challenges and significant changes of the twentieth century, agriculture is still very important in the state, generating revenue of about $4.5 billion a year. Yet this industry comes nowhere near the heft of the state's finance, real estate, and tourism industries. However, taking into consideration the state's historical and cultural identity, land use, and employment, the case for its prominence is legitimate.[122] In the Northeast, next to Pennsylvania, New York is the largest farming state. While agriculture in New York is small relative to larger farming states, it is nevertheless vital to the state's economy. And, despite its comparative position, nationwide, New York is exceeded only by California in the market value of direct-to-consumer sales, which includes farm stands, farmers' markets, and CSAs. Approximately one-fourth of the state's land is used for agricultural production, with 36,352 farms averaging 197 acres.[123] Within New York, the Hudson Valley is unique due to the small average size of its farms. Table 1 offers agricultural data to place the Hudson Valley and New York in comparative context nationally.

New York is among the top five states in the production of dairy products, apples, cabbage, grapes, cauliflower, tart cherries, pears, sweet corn, green peas, cucumbers, snap beans, and squash.[124] Yet the state's small producers struggle to compete with corporate agriculture to supply their produce to huge supermarket chains. Large retailers such as Walmart control produce prices nationwide and keep them low to attract consumers who will purchase many other items in their stores. Walmart dominates with one-third of the grocery market, and other big-box stores are heavily investing to expand their own grocery operations.[125] At the same time, these retailers are keeping an increasingly larger percentage of the profits from produce sales, while those of farmers are declining.[126] Another hand in farmers' pockets is that of the middleman transporter, whose costs are increasing with the rise of fuel prices in the past decade.

Now that farming is increasingly promoted as a central cultural component of tourism, maintaining the scenic presence of barns and working farms is critical.[127] This need is addressed by land conservation efforts on the part of townships, counties, and nonprofit organizations. Forty years ago, the region was dominated by large dairy farms and expansive grain operations. Those are largely gone now, and today agriculture tends to be less land-intensive. In addition, new technology has reduced the acreage needed for farm-

TABLE 1 Comparison of Farming among the Hudson Valley and Various States

	Hudson Valley[a]	New York	California	North Carolina	Florida	Texas
Acres of farmland	415,092	7,174,743	25,364,695	8,474,671	9,231,570	130,398,753
Number of farms	2,711	36,352	81,033	52,913	47,463	247,437
Average farm size (in acres)	143	197	313	160	195	527
Annual market value of ag products sold (x $1,000)	266,352[b]	4,418,634	33,885,064	10,313,628	7,785,228	21,001,074
Average annual market value of ag products sold by farm (in dollars)	98,032[b]	121,551	418,164	194,917	164,027	84,874
Annual market value of direct-to-consumer sales (x $1,000)	17,310	77,464	162,896	29,144	19,363	38,696

SOURCE: U.S. Department of Agriculture 2007 Census of Agriculture.

[a] Unless otherwise stated, data for the Hudson Valley include Putnam, Dutchess, Columbia, Orange, Ulster, and Greene counties.

[b] Data unavailable for Putnam County.

ing. For example, on one of the farms I visited, apple trees were planted only three feet apart, an economy of space that was unheard-of two decades ago. As a result, we can expect to see the continued alteration of the agricultural landscape as the land required for food production shrinks. "I've seen an increase in the number of new farms every year," observed a relatively new farmer, but he continued, "our impact on the land is another question because we don't take up the space that dairy farms do. Whether we'll be able to preserve open space in the Hudson Valley is a long shot."

Preserving open space and safeguarding working farms require different strategies that may be at odds with one another. A recurring topic among my interviewees was the high price of land, which is both a barrier to those who want to get into farming and an enticement for existing farm owners to sell to developers. One young farmer claimed that California's Napa Valley is the only other place where agricultural land is as expensive: "The soil is good here, but whether you want to put in a farm or build a house, it still goes for the same price." For this reason, conservation easements may keep farmland intact and prevent development, but they do not make land more available to farmers. Describing the challenges of farming on the "suburban frontier," one respondent bemoaned the estate buyers from New York City who were eager to preserve their open views, and who regularly leased sections of their land to garner agricultural discounts on their property taxes. Adamant that such practices would not save the region's farming, he declared, "We are creating a larger class of tenants and sharecroppers, and that's not right. I know that's our history in this country, but it doesn't get you to the best place for the land and food and agriculture." Some of these tenants are the new generation of younger farmers driven by the zeal of the food movement. And while leasing land does not carry the stigma that it did a century ago, the conundrum for the tenant farmer is the same as it once was: Do I invest in improving the land when my claim to it is so insecure?

TWO

The Workers

LABOR CONDITIONS, PATERNALISM, AND IMMIGRANT STORIES

CHARMED AND PERSUADED BY THE aesthetics of agrarianism, food writers sustain the belief that local agricultural activity is superior in almost every respect to the industrial food system. Indeed, in this regard "local" is sold as a commodity, one laden with assumptions that often go unexamined. We want to see our local/small/family farmers as much more than merely businesspeople. We expect them to produce a range of diverse crops, to either practice organic farming or the safer use of chemicals,[1] to have fewer animals than CAFOs (Concentrated Animal Feed Operations), and to allow their animals to roam freely and graze on grass. Farmers should also be land stewards who sustain regional foodsheds. As for the farm products, we know they are much fresher (picked yesterday!) and tastier than those we purchase at the supermarket. Given the constant reinforcement of these benefits, it is easy to see how consumers conflate local, alternative, sustainable, and fair as a compendium of virtues against the demonized factory farm. Consumers cannot be faulted for this; they are simply mimicking the attitudes of food movement leaders like Michael Pollan, who argues that there are two essential categories of farming: industrial and pastoral. The latter, considered the source of all virtue, is referred to with the interchangeable terms "'organic,' 'local,' 'biological,' and 'beyond organic.'"[2] But where does farm labor fit into this division of good and evil?

Food writers do discuss the wages and conditions of farmworkers, but usually only in relation to the industrial food system. For example, in her memoir of her family's year of eating local, Barbara Kingsolver addresses the poor pay and conditions of U.S. farmworkers and cites their national average annual income of $7,500.[3] Clearly, readers are intended to be upset by this and feel grateful that local farms offer a more highly paid alternative. This fits neatly with the idea that food ethics as championed by its leading proponents should include concern for the workers' conditions. A food ethic inclusive of labor concerns is also on display in the work of other popular food writers, such as Eric Schlosser, who excoriates the poor conditions of strawberry workers and meat-packing workers in industrial farms and meat-processing facilities.[4]

Since the general reader rarely gets to hear about labor conditions on local farms, what are the most important things to know? My analysis of Hudson Valley farms shows that many of the labor practices mimic those of the dominant food system. Additionally, these farms foster paternalism, a significant component of labor control that is unique to small-scale agriculture. There are in fact several overlapping factors that combine to create and sustain extreme vulnerability in the labor force. These factors help to explain why farmworkers largely accept, almost without questioning, the poor working and living conditions offered on the region's farms and why they do not organize to change their situations.

In this chapter I offer a picture of those regional conditions based upon the farmworkers' own words. Their testimony speaks to their immediate situation while invoking the historical and structural factors that have shaped and determined their livelihoods. To fill in the stories, I have added background on the regional labor market, a detailed analysis of work conditions, an account of the structure of this farm labor market, and a discussion of the intersection between immigration experiences and farmworkers' fears. I also explore the paternalistic price of proximity to explain common labor practices on regional farms.

BACKGROUND AND WORKER CONDITIONS

The average size of farms in the Hudson Valley is 143 acres, and in New York State it is 197 acres.[5] The region has a concentration of fruit, vegetable, and horticultural (FVH) farms, with primarily seasonal jobs that tend to offer the poorest wages and working conditions. These jobs were traditionally filled by migrant workers.[6] Many workers, including those I interviewed, were housed directly on the farms in "labor camps," a common term for housing created for farmworkers, such as trailers, barracks, or houses. The work performed by the farmworkers interviewed for this study includes a broad range of tasks: picking fruit, cutting vegetables, planting, packing, hauling boxes, sod landscaping, managing other workers, and preparing food. However, the vast majority of workers interviewed—nearly 90 percent—were primarily engaged in planting, harvesting, and packing.[7]

The farm labor market in the Hudson Valley, and in New York in general, has changed significantly since the early 1980s. Farmworkers today are less likely to be migrants, and increasing numbers settle year-round in New York.[8] Undocumented Latinos, mostly Mexicans, have largely replaced African American and Caribbean workers, who were the primary work force for the better part of the mid-twentieth century. In addition, workers today are largely undocumented, whereas before 1980 the majority were citizens or in possession of a green card. The Hudson Valley is unique in New York and among agriculture-intensive states for its lack of reliance on labor contractors, who act as middlemen for hiring, supervising, and paying workers.

Those who want to read about horrific labor exploitation will find examples in the pages that follow. My field research certainly turned up evidence of labor and human rights abuses, but my primary goal is not to offer an exposé of this wrongdoing. Instead, I aim to present a deeper, contextual analysis of the typical work experience of farmworkers on these local farms and the many factors that have shaped it. Violations of labor law were common, however, and I list some of them here to give a sense of their range and severity. In 1995, twenty-two duck workers were fired and evicted from their homes on a labor camp with the help of local police. A case brought by Farmworker Legal Services of New York on behalf of the workers led to their reinstatement in the jobs. In 1997, forty onion packers were unlawfully fired after requesting a pay raise. All except the worker spokesperson were hired back following a lawsuit. In 2002, four upstate New York farmworkers escaped from a labor camp where thirty workers were held in locked barracks at gunpoint and threatened with violence if they complained. The contractor overseeing the workers pleaded guilty to federal charges of forced labor and human trafficking. In 2003, cabbage workers were shot at while working in the fields. Two local teenagers with rifles were charged with the shooting. The charges were dropped in 2004 when the workers were out of state and not given appropriate notice to return in time for a hearing date. In 2004, a van accompanying a cross-state farmworker march with a "justice for farmworkers" banner in its window was shot at while it was parked overnight in a rural town.

Although such incidents are not everyday occurrences, news about them spreads fast and serves as a lesson to other workers. In this way, such abuses helped to discipline workers who may have thought about taking steps to address their own concerns. On one farm I visited twelve workers had recently been let go; the rumor around the camp was that the workers had complained about their housing. Tellingly, the replacement workers were thought to be housed in a local motel, which bred resentment among those who continued to reside in the farm's labor camp. For the workers I spoke to, whether the rumor was truth, exaggeration, or fabrication was less important than their belief that the scenario was likely. Similar unverified stories came to me from workers on most farms. It was apparent that both truths and half-truths fed into workers' perceptions about how they might be treated themselves and contributed to their reluctance to challenge their employers.

In a more typical scenario, a woman engaged in apple packing, which is supposed to be easier than field work, told me, "You are dead by the end of the day; your arms and your feet ache because of standing all day." One field hand told me he thought dogs were treated better than he was; he then worried about telling me too much. Caveats like this were not uncommon, and many workers said they were reluctant to share stories about their work conditions. They commonly used phrases such as "I better not say" and expressed fear of getting in trouble. Repeatedly, however, I was informed that their work was very difficult, and the vast majority detailed how they suffered in particular from adverse weather conditions, as most were expected to work in both the rain and extreme heat. In terms of overall treatment, workers on about one-third of the farms I surveyed (32.2 percent) reported that they were not treated respectfully. Not a few described their harrowing treatment and compared their working conditions with those of slaves. Yet these were workers who returned year after year to toil on behalf of their disrespectful employers because job security and a guaranteed source of income outweighed all other concerns.

The average total annual income for workers surveyed was $8,163 in 2001 and $8,078 in 2002. These figures include wages from Hudson Valley farm work[9] and, for 36 percent of respondents, additional income.[10] This very low income level was well below the 2002 U.S. Federal Poverty Guidelines for a family of three ($15,020).[11] When I interviewed workers again in 2008 and 2009, their wages were slightly higher. These findings correspond to other studies, such as the 2011 report on U.S. food workers, which found that 93 percent of agricultural workers interviewed earned a poverty or subminimum wage.[12] While the federal minimum wage in 2002 was $5.15 an hour, the average hourly rate for those interviewed was $6.92, with almost half earning between $6 and $7 an hour.[13] While 84 percent of workers expressed that they should be paid more, I was surprised by the modest amount they desired, on average $8.53 an hour, with a range of $6.50 to $18.00. Farmworkers' indigence is perpetuated by the fact that agricultural employers are held to different labor standards than other employers; for instance, growers in New York State are exempt from paying overtime. In addition, farm work is seasonal, and so even if it is well paid in season, the net income over the course of a year does not compare well with other low-skilled jobs in the United States. With regard to pay, many workers told me that it simply was not possible to be paid more, nor were many comfortable asking for a raise. Workers and employers did tell me of situations where workers did ask for increased pay, and some received it, but I heard just as frequently that those who had asked for raises were told to leave if they were not satisfied.

Inconsistency was a perennial feature of these agricultural jobs. Workers reported fluctuations in the hours and days they worked from week to week. Although this is due to the fact that farm work is heavily dependent upon both seasonal demands and weather, such unpredictability also leads to economic instability. When asked how many hours they worked, more than half reported that their hours varied greatly, but the rough average for a typical week was forty-nine hours. Workers engaged in apple packing told me they regularly logged forty hours a week, whereas harvesters reported eighty- to ninety-hour workweeks for several months in a row. They told me they often worked from six in the morning until 10 or 11 at night, with a midday break to return home for lunch but little time for dinner, which was consumed in the fields. Not entirely believing this, I drove past the farm on several occasions to verify these work hours and indeed saw harvesters leaving the fields after dark and continuing to work in the packing house past 10 p.m. The farm's bookkeeper informed me that hours above fifty or so were not on the books and would never appear in official data. Overall, almost half of those I interviewed reported regularly working more than five days a week.[14] Many workers—both field workers and those working as packers—reported being yelled at or rushed when on break while the American-born members of the labor force could take breaks whenever they wanted. There was a consensus that more breaks should be allowed when working in extreme heat.

When workers compared their situations to those of their native-born counterparts with manufacturing jobs or working as mechanics, they complained that they did not receive paid sick days. None of the workers I met received paid sick days. Although the consensus was that employers were generally understanding when it came to employees needing time off when they were sick or had a family emergency, workers owned that they rarely missed work, and several mentioned that their boss would be annoyed or angry when anyone was too sick to show up. Again and again I heard about those who went to work with a fever or a severe cold. This corresponds with data that shows that 53 percent of U.S. food workers go to work ill.[15] Workers also compared their situations to those of their white, U.S.-born coworkers, most of whom did not harvest or pack but worked as mechanics or tractor drivers or in similar positions. An apple packer mentioned that whites could go home without a problem if they were feeling even a little sick. One worker told me he lost his job when he hurt his back while an American coworker was allowed to return and take it easy after an injury. Guest workers reported that their peers had been sent to their home country after being injured on the job and had to fight to be reimbursed for medical expenses. Several workers reported that their boss paid for hospital bills only to deduct the money from their paychecks. No one was provided with health insurance, although low-cost clinics were available for farmworkers. Pesticide exposure was a widespread concern. One family described how they tried to stay in their trailer and kept their children inside for two days straight in order to avoid exposure after they noticed that the pesticides being sprayed in the fields made them feel weak.

Almost all the workers I met lived in farmer-provided housing, including some who lived rent-free, since, as they told me, it was very difficult to pay for nonfarm rental housing on their wages. In the labor camps I visited, the housing varied widely, ranging from trailers and cement-block barracks to large rooming houses, which were sometimes divided into apartments. Some were well maintained and offered ample space for dwellers, but many were run-down and crowded. The trailer of one guest worker was meticulously clean and boasted many comforts. In contrast, on a neighboring farm, a small two-bedroom trailer in disrepair housed eight workers and had four bare mattresses piled up in the living room. Some housing was entirely unadorned, with the most personal items on display the workers' discarded boots, while others had many personal touches, including posters, photographs, party favors from baptisms and *quinceañeras*, and an array of cooking utensils. In another trailer a worker had amassed an abundance of plants and a fish tank.

Securing adequate accommodations was an acute problem for most of my interviewees. Although 60 percent saw no need to change anything about their housing, this was not an indication that they had clean and well-kept accommodations. Rather, it reflected their willingness to tolerate poor housing conditions: windows replaced with cardboard, missing screens, broken outdoor lights, mold in bathrooms and kitchens, peeling or missing flooring, and leaky roofs. Of course, their reluctance to complain could also be related to workers' expectations when they compared their New York housing to that in their home countries, but there, proper insulation from the cold was not a concern. In many instances I saw clearly inadequate shelter in which the occupants nonetheless insisted that they would "change nothing." In one case, where eight workers shared a room and slept on thin, bare, filthy mattresses, not one saw fit to complain. In another instance, workers did not even report that they lacked mattresses to sleep on, a detail that only surfaced after I pursued the topic. Apparently they were so afraid to ask their employer for beds that they opted to sleep on the floor. Reluctance to demand such basic necessities displayed the psychology of extreme compliance and sacrifice that is shaped by their situation.

Nonetheless, I encountered no end of ambivalence about whether their sacrifices were worth the wages and whether the economic benefits outweighed the social deficits. One apple packer attested to what he had gained and lost while employed as a farmhand. He explained, "I have built a house; well, I should say a mansion, because I made it to my liking. It cost me a lot of money, a lot of money. But it also cost me separation from my family, which is the thing I love the most." He was unable to live with his family in this mansion back in Mexico as long as he needed to work in the United States to earn money for their upkeep. Another pointed to the acute economic difference between the United States and their homes. "You can buy a blender here without even thinking about it, but in my country you would need to save for two or three weeks. Here, in one week, you can even save enough to buy a TV." (Not every interviewee reported having such surplus income since they sent such a large percentage of their wages home.) His coworker was equally conflicted, saying, "It's complicated. I know that I am better economically here, but we are so lonely, it is depressing," and he continued by describing feeling alienated from American culture.

More challenging than the work, without a doubt, was the workers' general sense of loneliness and separation from their families. Gloria, a twenty-two-year-old Guatemalan woman broke into tears while recalling how much she missed her home. She spoke to her mother often by telephone, but she said that she never related her sadness or complained about the work. Like others who downplayed their hardship, Gloria's overriding goal was to optimize her income even as she was acutely aware of her meager earning potential.[16] No one harbored illusions about the United States as a land of opportunity. As a former farmhand summarized it, "Everyone talks about the American dream, but for us it's more of a nightmare." When asked if they would want their children to follow them into farm work, workers answered with a resounding "no" in almost every case, and many remarked on the suffering they had endured. One apple picker asserted conclusively, "There is no future in the fields; there is only money." And it did offer that. Ninety-five percent of the workers I interviewed sent money home, and two-thirds did so on a monthly basis or even more frequently. The average remittance was $513 a month, a figure representing roughly one-half of an average worker's monthly take-home pay.

STRUCTURE OF JOBS

Considering the nature of the tasks, the wages, the hours, and the seasonality of farm work, it should not be surprising that historically this sector has been the reserve of immigrants, whether European indentured servants, chattel slaves from Africa, refugees from economic and political violence, or immigrants seeking a better opportunity. Growers have also hired citizen workers, who have been marginalized socially and economically because of the poverty and institutionalized racism that comes with the job. Though the payrolls of some of the region's farms include a few highly skilled workers and perhaps a farm manager employed at higher salaries, the vast majority of farmworkers have generally been wage laborers, employed for seasonal tasks such as harvesting. Consequently, the evolution of the farmworker job category has been shaped by the vulnerability of those who were hired to its ranks. However, to consider the reproduction of farm labor inequality as based solely on the prior poverty and the desperation of its prospective work force would be to ignore other factors, which include the labor laws governing farm work, the short-term nature of the work, rural isolation, the impact of guest worker programs, and the absence of union organization.

Excluded from Labor Laws

Farmworkers in most states, including New York, do not enjoy the same legal protections as most other workers.[17] The exemption of the agricultural industry from New Deal labor legislation relegated farmworkers to a lower, separate category of laborer and guaranteed that they would be much more vulnerable to exploitation than their waged counterparts in other industries. They were excluded from collective bargaining protections, overtime pay, and the right to a day of rest. Until 1999, New York's farmhands were not covered by the state minimum wage but were subject to a different, lower standard. Moreover, child farmworkers do not have the same protections as children in other industries.[18]

In addition, smaller farms are held to lower labor standards. If a grower in New York has fewer than five workers on the clock, a portable toilet need not be supplied as long as workers have access to transportation to sanitation facilities. The rationale for these exceptions is that the costs of higher standards would put too large a financial burden on small farmers. The upshot is that the farms that are the most idealized by those in the food movement are not required to offer the same labor protections as larger enterprises in the industrial agricultural system.

The exemption of agricultural workers from labor laws dates to an era when the southern Democrats' lock on national electoral politics was unassailable, resulting in white supremacist politics that assured the perpetuation of a low-wage, southern, black work force.[19] The National Labor Relations Act (NLRA) of 1935, which established certain collective bargaining protections (such as requiring employers to recognize and negotiate with unions), did not include domestic and agricultural workers, because their work forces were predominantly black.[20] These workers were also left out of the 1938 Fair Labor Standards Act, which instituted workplace practices such as the forty-hour workweek and a federal minimum wage.[21] Orphaned from collective bargaining protections, farmworkers were unable to organize to counter the ill effects of the other legislative exclusions, including wage protections. The continuing legal neglect of agricultural laborers was a direct result of their lack of political power and the contrasting clout of farmers' organizations.

It is important for those interested in ethical eating to understand that it is not only the violation of labor laws, but also, and perhaps more importantly, the institutional marginalization of agricultural workers—including the denial of basic labor and human rights—that reproduces their inequality in the workplace and confines them to substandard working conditions. Following the letter of the law can still result in extreme exploitation. Consider the field workers I met who logged ninety hours a week at fifty cents above the minimum wage, or the dairy worker who milked cows for sixteen hours a day, six days a week, at $8 an hour. These were normal work regimes for months at a time, not the result of the pressures of harvest time for a few weeks. Workers who raised the ducks for the prized delicacy foie gras told me that they clocked in for three four-hour shifts a day with four hours off in between; this schedule was repeated twenty-four hours a day for many weeks at a time. The result was a cycle of eighty-four-hour workweeks, with no time for prolonged sleep, and all of it at a flat pay rate without overtime compensation. Foie gras producers insisted that they needed one worker who was dedicated to the same ducks throughout the feeding cycle to limit the stress on the animal and assure the quality of the final product. The stress on the workers, however, who might go for months without more than a four-hour break and three and a half hours of sleep at a time, was considered secondary to the health of the ducks.

The plight of these workers was featured in a 2009 article by *New York Times* columnist Bob Herbert, who quoted the owner of Hudson Valley Foie Gras as saying about his employees, "This notion that they need to rest is completely futile. They don't like to rest. They want to work seven days." Herbert reported that a New York Farm Bureau spokesperson concurred.[22] Among the flood of readers' comments was one particularly biting response: "One can't help observing how the owner and the Bureau have conspired to render silent the mouths of both the geese and their feeders—one genetically, the other by coercion."[23]

Rural Isolation

Workers housed on or near farms generally found themselves in a state of isolation—both geographic and social—because they were confined to hidden labor camps without much access to transportation. Cloistered from the local communities, they were unlikely to have neighbors aside from the farmer's family. Studying the spatial distribution of Latinos in the United States, one scholar noted a trend in rural areas: Latinos are clustered with other Latinos on a local basis, but they are segregated from other racial and ethnic groups.[24] In addition, rural areas that have seen a recent surge in Latino populations do not have the well-entrenched support networks, services, and advocates found in traditional immigrant gateway cities.[25] The Latinos that I interviewed reported that the families and friends who helped them find jobs in the United States did offer them a community of peers, but this semblance of a social network also reinforced their isolation by separating them from their host communities.

Workers' isolation was exacerbated by their lack of access to transportation, a particular obstacle in rural locations where public transportation is virtually nonexistent. Moreover, the vast majority were unable to obtain a driver's license legally due to their immigration status. Eighty-five percent of those in my study relied on their friends, family, employer, or a paid ride for their basic transportation needs. It was quite common for a grower or chief laborer to bring workers into the local town to shop for food, but this rarely happened more than once a week; weekly transportation is required on farms that hire H-2A guest workers but not on others. "A person who doesn't have a car is a person who doesn't have feet," according to two Columbia County migrant workers I spoke to. Their possession of a vehicle allowed them to travel from farm to farm along the East Coast each year, but they joked, "We only have licenses from God." Those who do have cars are pressured not to use them. Employers with whom I spoke had advised their workers to stay as close to the labor camp as possible. Although farmers were looking to protect their employees (and themselves) with this advice, it had the effect of further isolating new immigrants from local communities. One field hand reported that some states, such as Florida, would issue a license plate when a passport is presented as identification, but this was not allowed in New York. Despite the advantages they conferred, out-of-state plates attracted too much attention in the small towns of the Hudson Valley. This was certainly true for some of the farmhands I met, whose license plates were stolen from their cars and the inflammatory slogan "Immigrants Leave!" scrawled on their car windows.[26] When a repeat of this incident occurred, they contacted the police and found themselves under suspicion and fingerprinted—in essence, criminalized—even though they were reporting being victims of a crime.

Guest Workers: Taught to Be Quiet

New York has been employing foreign guest workers since World War II, primarily from Jamaica, but more recently from Mexico.[27] The state currently admits about three thousand foreign guest workers a year to work on farms. Before receiving approval to hire guest workers, an employer must demonstrate that no viable domestic workers are available; they do so by posting job listings in newspapers and on Internet job sites. Guest workers can be in the country for up to a year. To prevent unfair competition with domestic workers, the federal government sets an Adverse Effect Wage Rate (AEWR), which is higher than the federal or state minimum wage, often by several dollars; in 2012 the hourly AEWR for New York was $10.56 for field workers.[28]

Despite significant reforms in the structure of these programs, the guest workers I interviewed were in a particularly confined situation that has been described as state-sponsored quiescent labor,[29] and they had limited options for remedying their problems. Legally in the country on work visas that were tied to a specific workplace, they were not able to terminate employment with the same ease as undocumented workers, and their retention in contracted work largely depended on their employers' positive assessments. Consequently, they had less incentive to complain about their work situations. Moreover, the economic and supervisory relationship between employer and guest worker was more complicated than that between employer and undocumented worker. The country sending the guest worker has a high stake in providing "good" workers—defined by my interviewees as those who did not complain or ask questions—since the bureaucratic apparatus for supplying and monitoring workers generates revenue and provides jobs in the home country. Countries sending workers also have liaisons in the United States to monitor and help guest workers, yet my interviewees reported how the monitors had threatened not to readmit workers to the program by reminding them that others were standing in line for the opportunities. In addition, workers who became ill or injured, who asked for different job assignments, or who complained were likely to be marked as "problematic." A forty-something Jamaican guest worker who had been working in the United States since 1980 summarized how he and his coworkers dealt with problems on the job: "Sometimes we don't. We are taught to be quiet."[30]

Among all my interviewees, Jamaican guest workers were by far the most cautious. Their responses were often as brief as possible, and some divulged that they could not comment on some of my questions because of fear of their bosses. The most common refrain among guest workers was that they had to follow the rules and do what they were told, whether it was in the contract or not. As Watson, a former guest worker, explained, "Most of the farmworkers that are on the contract don't want to speak to people." He and other former guest workers were much more open than those currently enrolled in the program. Indeed, during one interview Watson, who is now a U.S. citizen and employed on the same farm, kept interrupting to urge a guest worker to tell the truth and elaborate on his answers. As he later explained, "If the guest worker breaks the contract, they lose the job. Whereas I can tell the boss, 'Do it yourself.' Boss and I can get pissed at each other and later we break it down and we compromise. If the boss gets on my nerves, we joke. Guest workers can't do this." For Watson, the ability to disagree with the boss was a matter of basic respect. Another former guest worker recounted that he had not been called back after he had questioned his benefits related to missing work for a week due to an injury. "If they find you are too sensible or too smart," he observed, "you don't have any chance on the contract, so you have to cover up your wiseness and your sensibility." Such comments reflected the pressure on guest workers to maintain good conduct, but they also spoke to the institutionalized modes of controlling and disciplining the workers' emotions, intelligence, and opportunities to communicate.

Union Neglect

One structural factor that inhibits collective action among the state's farmworkers is the absence of union organizing efforts. Despite the high visibility of West Coast organizing efforts by the United Farm Workers of America, as well as the more recent Coalition of Immokalee Workers in Florida, the primary channel for low-wage worker dissent—the labor union—has been mostly inaccessible to farmworkers.[31] From the union perspective, organizing such workers is logistically difficult and expensive, particularly when they do not even have collective bargaining protections. The strategic obstacles to organizing farmworkers were intensified in states like New York, where the growing season is short, farms hire comparatively few workers, and geographic dispersion hinders worker solidarity. Even where high concentrations of workers and longer growing seasons exist, such as in California, unionization was deterred due to a surplus of agricultural workers, fiercely antiunion growers, and the influence of coordinated farm owners on rural representatives in the state legislature. This created a significant twentieth-century division between field and factory, with agricultural workers trapped in a state of extreme disadvantage, and urban industrial workers' conditions and wages improving over time. The disparity only intensified as the gap between them in skill level, job security, union attention, organizing successes, and legal protections grew. Until the mid-twentieth century, most unions identified agricultural workers as unorganizable.[32]

PATERNALISM: THE PRICE OF PROXIMITY

For the most part, the circumstances of Hudson Valley farmworkers directly mirror the circumstances of low-wage immigrant workers around the United

States, including those on corporate industrial commodity farms. However, one aspect that is different is the paternalism that thrives on the small family farm as a primary mode of worker control. I argue that this paternalism stems from the intimacy between workers and employers on smaller farms, resulting in a "price of proximity" for the farmhands. Workplace paternalism by employers can be understood as an intimate but extremely hierarchical relationship in which the employer's control extends into workers' everyday lives, affecting even their personal and recreational habits.[33] In a paternalistic setting, employers typically extend benefits to workers in return for good behavior and loyalty.[34] Such benefits revolve around individuated relationships that address workers' material and psychological needs.[35] On the small family farms I surveyed, the system of paternalism was relatively complex: there were varying degrees of benefits, which resulted in different levels of involvement in and control over worker habits and behavior.

These benefits extended far beyond the work contract and could include material perks such as free farm products, as well as occasional help, for example, in resolving a problem facing workers' children. More significant assistance might include helping a worker secure a green card or offering protection from immigration officials. The benefits to workers were irremediably disproportionate to what they might provide for themselves, meaning that employees were not in a position to reciprocate, and the disparity in power that is bolstered by such benefits helped to create and reinforce quiescence within the work force.[36] Workers are encouraged to develop an image of a benevolent boss and reciprocate in the only way they can: with good behavior in the workplace. This results in worker loyalty and ultimately minimizes labor costs.[37] Conversely, workers who do not deliver good behavior might find themselves ineligible for such benefits.

The most prominent examples of paternalistic employer relations in the United States date to the late nineteenth and early twentieth centuries, when methods for controlling free black workers, the poor, and new immigrants evolved on southern farms and in southern mill towns. These specific systems of planter and industrial paternalism no longer exist. After World War II, machines not only replaced workers, but they also standardized agricultural production in a way that served to discipline workers.[38] Mill town work culture was dismantled due to the pressures of modernization, and was replaced by rationalized industrial practices that could better compete in the growing international market.[39] Yet remnants of paternalism are still in place and have taken on a more contemporary form: while planter control was based on racialized inequality, and that of the mill owners was founded on workers' poverty, today legal status plays the most important role in reinforcing the hierarchies between bosses and workers.

Additional factors feed into paternalism. Because this kind of relationship draws on their personal knowledge about individual workers, employers need to know the first names of their employees and have some intimacy with the details of their lives.[40] Workers must perceive the employer as having authority or esteem beyond just ownership of the means of production.[41] For Hudson Valley farmworkers, status factors such as national origin, citizenship status, social class, level of formal education, language spoken, and skin tone played a determining role in the dynamics between workers and their white employers. Inequality resulting from these factors was further exacerbated by the disparity in levels of "membership" workers could hold in both their communities and the political system as a whole; furthermore, workers' limitations in these areas often directly coincided with race in mostly white rural towns.[42] Paternalism typically thrives when the jobs on offer are a significant step up from what workers previously held. In this sense, the nineteenth-century mill owner would be perceived as rescuing his employees from more vulnerable positions and offering a "path toward economic and social stability."[43] Similarly, today's Hudson Valley farm jobs offer relatively valuable income opportunities to poor, low-skilled immigrant workers, most of whom have limited formal education, to support their families in a way not feasible if they had stayed in their home countries.

Conditions for paternalism were especially ripe on Hudson Valley farms. First, hand labor is the norm, and so the workplace entails close management of manual labor tasks. The harvesting of easily bruised fruits and vegetables is traditionally difficult to mechanize, and when machines do exist the price can be beyond the budget of the region's farms. Nor are there one-size-fits-all harvesters suitable for farms with diverse crops. Contrast a farm with diverse crops, for example, with a monocrop farm in California's Central Valley that makes use of mechanical harvesters to help process six million pounds of carrots daily.[44] Second, grower paternalism relies on intimate personal relationships. All of the farmers I interviewed were directly involved in managing their workers and knew them all individually. In the Hudson Valley there is an absence of farm labor contractors, or middlemen who mediate labor relations for farmers, including doling out responsibilities and pay to workers. They are more common on the larger farms in Western New York and on industrial commodity farms around the United States. Without contractors or middle management, the farmers give more personal attention to employees and can be more sympathetic to their daily situations.[45] Third, as the business literature on small firms highlights, paternalistic labor relations are common in family enterprises.[46] Small family businesses usually do not have middle management, and on farms in particular the vast majority of employees are wage laborers.[47]

In the Hudson Valley, which is increasingly a regional agricultural system that feeds local food markets, the intimacy of employer-employee relationships creates for the workers what I call a "price of proximity" that makes them vulnerable to labor control. The consequence of this price is the reinforcement of paternalistic power disparities between workers and farmers and a labor regime that serves to deter collective action. Is this the same personal attention in the workplace that is promoted by food writers as integral to the virtues of local food? In many respects, yes, and it bolsters a type of labor control that is less likely to exist when the farmers do not have a direct role in supervising workers. The hands-on contact with and care in producing food that is the stock-in-trade of the farmers' market and the foodie magazines translates directly into labor paternalism in the fields.

Since place functions to communicate social identity, control over employee housing is another key element of the labor regime. When an employer is in a position to regulate both their employees' work and home, paternalism is an even stronger force. On the one hand, offering free housing to workers, permitting their nonworking family members to live on the farm, and allowing workers to use grower-owned vehicles for transportation can be perceived as beneficial. On the other hand, such practices directly reinforce power hierarchies, impede collective dissent and action, and generate new kinds of inequality.[48]

For farmworkers who lived in labor camps there was an implicit and explicit understanding of employer involvement, not only in the employees' work lives, but in their personal and recreational habits as well. Social life usually took place on the farm premises, where, on Friday or Saturday evenings, it was common to see workers gathered around cases of beer piled high. Soccer matches were an occasion for socializing, and more than one farm I visited had a regulation-size soccer field. The intimacy established between workers and employers in this environment creates a relationship in which the latter has much more control over workers than they would if the workers lived off the farm. Like live-in domestic workers, employees were on call at all times.

The proximity of work and home also has radical consequences for workers' identity and for their economic and social opportunities. Labor camps highlight the power structure of the agricultural industry. Located on farm property, the camps were highly segregated spaces that rendered the outside world all but inaccessible. The degree of isolation is not unlike the experiences of those in inner-city ghettos who have little contact with other social groups.[49] In addition, most farmworkers were newcomers to the United States with significant language barriers, and so they were cautious about interacting with the public. Employers often discouraged their workers from taking trips off the farm in the name of protecting them, but this only reinforced their isolation from local communities and solidified their dependence. Workers' missed opportunities were extensive, ranging from access to local amenities and knowledge that could be gained from media and community bulletin boards to social capital from making contact with others. In addition, farmworkers received little local media coverage, especially in comparison with the farmers who employed them. Self-empowerment through networking, collective action opportunities, or political action was generally closed off for those who could not move freely beyond the orbit of the camps. The wife of one farmworker told me that she left the farm only to clean houses and visit the grocery store, and her children left only to attend school. Although her husband owned a pickup truck that he drove around the farm, he rarely took his family anywhere because he did not have a driver's license and did not want to put his family at risk.

In a more subtle manner, housing reinforces social rank through the quality of the home interior and exterior[50] and through landscaping in the neighborhood[51] or urban setting.[52] The hierarchy that existed between growers and workers was not only based on their respective status as boss/worker and owner/tenant, but it was also reinforced by poverty (the average hourly rate for those interviewed was $6.92), legal status (71 percent were undocumented and 21 percent were guest workers), race (99 percent were foreign born), education (the average grade completed was sixth), literacy (80 percent reported they could read and write but demonstrated very low literacy), and proficiency in English (for Latin American–born workers, the average self-reported English-language level was 1.2 on a scale of 0 to 5).

In contrast with their workers, New York growers were usually white, land-owning, educated citizens. The attendant power differential between the two groups was embodied in their respective domiciles. For example, one orchardist's two-story brick house, which housed her single family, was located close to farmworkers' trailers, some of which housed up to eight workers in two bedrooms. A material symbol of their power, the employers' homes, although not on the scale of southern planters' mansions, were almost always much larger and of a much superior quality to what was provided for workers. Farmers also understood that labor camps could sequester workers from a community that does not want to see them. "Having them down the lane keeps them out of people's faces," observed a Dutchess County farmer, who added, "it keeps them quiet." Trying to picture his workers living on the town's main street, he remarked, "I think people would frown upon it."

For employees whose housing was tied to their jobs, labor control also extended to nonwork locations and hours. The relationship embodied in the labor camp was one of dependence on the employer, not only for shelter, but also for safety. This holds true for most workers whose housing is tied to their jobs, but it is heightened when workers' families live with them in grower-owned housing. Almost a third of my interviewees had wives who lived with them, and more than a fifth lived with their children. Although a grower may be driven by moral sentiment in permitting families to live in the labor camp, for the worker, the need to keep his job affected the housing and livelihood of his entire family, including child care or schooling. So although families were somewhat more expensive and inconvenient for employers to accommodate, the workers' dependence on the farmers resulting from their family obligations was exploited in a way that was not the case for single males. Though most of my interviewees were grateful for free or low-cost housing, they acknowledged that the circumstances made it difficult for them to consider living off the farm, no matter how uncomfortable they might be. The prospect of paying market-priced rent was a huge deterrent, as was the difficulty of obtaining references, credit, or funds for a security deposit.

Although employers were required to supply housing and transportation for guest workers—the most basic subsistence benefits offered through paternalism—the same was not true for other workers, for whom such benefits were rarely formalized in writing. The majority of my interviewees lived in free housing, but most of these arrangements were casual and could change at a moment's notice. On one farm I visited the workers' rent had been increased midseason, while on others workers had been required to pay for benefits that were previously free, such as electricity, heat, and telephone and television services, with little warning of the change and no opportunity to contest the additional costs. Another benefit that was not formalized was ready access to farm produce, though, again, this perk could be withdrawn at the whim of the employer. Much rarer was the farmer who had allowed workers to create a chapel, which was visited weekly by a priest.

Employer-employee intimacy meant that growers often helped workers with their problems, most of them related to money or other resource shortages. This assistance comprised the second, more significant layer of benefits, beyond basic subsistence, that could be supplied through paternalism. Several workers told me that they had received advances on their salaries to help with family emergencies. One farmer paid the round-trip airfare for his Jamaican guest worker to return home for his mother's funeral (the worker reported that the contract had no system in place for such family issues, which he thought was unfair). Another boss helped secure a place and a scholarship at a private school for his farmworkers' child who was being bullied in the local public school. Other examples included helping workers obtain the cheapest flights home, responding to a complaint from a local bar about drunken workers, developing an insider relationship with local police to ensure immunity for minor traffic infractions, and holding a job for a favored worker when he returned home for a year. All of these actions by Hudson Valley farmers are examples of the paternalist employer who is in a position to "help," and there was no shortage of testimony from employees about how they felt indebted for the assistance.

An even higher level of benefits was promised to some of the undocumented workers. Many had been promised assistance in securing visas. I heard of one case in which a green card was successfully procured and another in which previously undocumented workers returned as guest workers. News about these cases traveled far and wide, and they gave (mostly false) hope to other undocumented workers. Employees also saw farmers as their protectors, especially in the face of hostile community members. In this regard, being cloistered in the labor camp was perceived as a clear advantage. Another significant benefit that a farmer mentioned was the possibility of selling the farm to some of his favored workers, a proposal that he had discussed with them. Offers and protections such as these comprised a third layer of paternalistic benefits.

Paternalism also allowed farmers to manipulate the loyalty of their workers. On one Hudson Valley farm, the grower was in the red and uncertain he would survive the season. He had discussed the situation with the workers, who agreed to defer receiving wages until the farm was able to support them. This particular farm owner otherwise seemed to be generous in his treatment of these year-round workers, paying year-end bonuses, for example. However, he crossed a line (and violated the law) by asking wage laborers to forgo their earnings for the good of the enterprise. This conundrum was further illustrated by a full-time orchard worker who, when I asked how he would resolve a particular problem with the boss, said, "Honestly, I wouldn't know what to do, because the boss is a good person and gives us a home and doesn't charge us rent."

Paternalism, of course, is based on a system of punishments as well as benefits. Workers told me that those who were not wholly compliant with management might be offered fewer hours, assigned to more arduous tasks, or be denied raises when others received them. I was told that when pay was dependent on piecework rather than an hourly rate, growers penalized less favored workers by giving them less productive tasks, for example by assigning them apple trees that had already been picked over a few times.

When a grower went the extra mile for an employee, was it an act of goodwill or an expression of labor control? In reality, there was no way of separating these intentions. Many farmers cared about their workers on a personal level, and they recounted specific examples of how they had tried to improve the quality of life of their workers, especially when they faced personal hardships. Farmers appeared to be earnestly proud of the opportunities their jobs offered to those who were so poor. I had no reason to doubt these sentiments, any more than the workers themselves who believed in the authenticity of their employers' concern. But this species of benevolence is inseparable from the exercise of labor control by nature of the employment relationship, and, when all is said and done, paternalism serves to benefit the farmers' businesses. Take, for example, the farmer who was aware that community members looked at his workers with suspicion. He explained that he coached his workers to drive slowly and be respectful in stores and urged them not to cause trouble. As he put it to me, "I've got to be careful. I don't want them to have any problems, and I don't need to lose a work force." In this comment, his concern for both his workers and his business was equal and almost indistinct.

Arguably, the key to understanding labor paternalism is analyzing why workers seem to consent to such relationships. One of the arguments of this book is that farmworker inequality is a long-standing structural component of the agricultural sector. Growers reinforced this inequality through hiring practices when they favored vulnerable workers who would not complain. A second factor is that workers had already risked so much to find these jobs. Consider how much workers had at stake. One Salvadoran woman borrowed $10,000 to get to New York in 2000; a Guatemalan reported the transit cost at \$5,000 in 1999, and by 2008 it was \$7,000. In 2007 the going price for the short trip across the Mexico–United States border was between \$1,500 and \$3,000, not including the next leg of travel to New York. Many of these immigrants had borrowed, scrimped, and sacrificed to get to New York in the first place, and as a result they felt responsible not only for respecting the investment they had made, but also for promptly paying off debts to family members. In addition, they sent much-needed remittances home to their families for food, clothes, homes, and education. Family and friends from home were able to see how far the remittances went, and so they were motivated to provide the same. The outcome was workers recruiting others for the same jobs, or what Charles Tilly called "opportunity hoarding."[53]

Finally, paternalism itself complicated the employment relationship for workers, because the perceived inequality is mitigated by the caring acts of employers.[54] Paternalism could be interpreted as kindness or even affection. As long as workers were seen to benefit, such a relationship was legitimized, but it clearly masked inherent conflicts about their labor conditions.[55] "I am obligated to do what the boss says," remarked an Ulster County apple worker, "so I work the hours he says, because he is giving me everything." In more extreme instances workers identified with their employers and adopted their boss's viewpoint and sense of self-interest.[56] Explaining why he would not want to join a union, one worker explained that it "might hurt the boss."

Growers whom I interviewed found it easy to justify paying no overtime by pointing out that workers had free or low-cost housing and access to farm products. Many told me that they knew from personal interactions that their workers were, in fact, satisfied with their jobs. On the other hand, some of their comments also displayed their awareness that such benefits, when extended to vulnerable and dependent workers, were a useful form of labor discipline. In particular, the existence of the labor camp highlights the lack of workers' autonomy and brings into relief not only the distinctions in social rank between employee and employer, but also the extent to which workers' lives (and sometimes those of their families) are entirely dependent on their good behavior in the workplace.

NOT JUST ANOTHER IMMIGRANT STORY

Paternalism is one among many mutually reinforcing factors that deny those who grow and harvest our food the opportunity to voice their concerns about poor working conditions. The lack of protective labor laws, the seasonal nature of the work, rural isolation and lack of community, and the structure of the guest worker programs all inhibit workers from acting to change their situations. An important cause *and* effect of these factors is the lack of political influence wielded by workers themselves. This was true for twentieth-century citizen farmworkers, particularly African Americans and migrants, and the problem is exacerbated today by the fact that so many of New York's agricultural workers are noncitizens.[57] Only 8 percent of the workers I interviewed were citizens or held a green card; 71 percent were undocumented; and the remaining 21 percent were guest workers. Although undocumented workers have populated the country's farms for the better part of the twentieth century, their concentration in the Hudson Valley and New York agricultural labor market is relatively new, dating to the mid-1980s.

The immigrant composition of the farm work force directly informs the workplace. Virtually all of the workers I interviewed were foreign-born. The most common places of birth were Mexico (two-thirds) and Jamaica (one-fifth), and roughly three-quarters of the workers were from Latin America. Workers' anxiety about their undocumented legal status combined with their aspirations to return to their home countries permanently influenced their decision to accept their situations without complaint. In particular, workers' plans to return to their home countries allowed them to rationalize their situations through habitual comparison of their current lives with their lives there, rather than with those of other U.S. workers. In addition, poor English-language skills, low levels of formal education, and a lack of job skills constrained workers' job opportunities. In general, these factors led to a willingness for self-sacrifice that limited making plans for the future.

During one typical farmworker family gathering after work, brothers-in-law Javier and Raúl, who belonged to the 22 percent of workers I met who traveled along the East Coast migrant stream every year, described their experiences. As we talked, the sounds and smells of dinner preparations permeated the room, a baby cried, a son required a ride home from sports practice, and a teenager lolled on the couch. We were seated at the kitchen table, which was clearly too small to accommodate the five adults and seven children who lived in this four-bedroom trailer. As Javier discussed the two families' annual itineraries, which took them from Florida oranges to New Jersey blueberries to New York apples and back again, Sonya, his teenage daughter, wrapped her arms around his neck. As Javier described his wages and hours—$7.50 an hour for packing apples while Raúl earned $8.50 in the fields—she massaged his face. Sonya and her three siblings were his reason for stringing together these workplaces. It was for their well-being and their futures that he endured the traveling, tolerated the low wages, and suffered the dirt that clung to his rough skin. He had adjusted to the cold New York autumns, but he had not adapted well to the daily fear of being deported and separated from his family. As he carefully removed Sonya's fingers from his cheeks and chin, he leaned in toward the table and explained, "We only ask that you let us work. I like being in this country, but not in the shadows, like they say here. I'd like to drive around without any fear, and go to Mexico and back. Obviously I'm not a citizen, but I think we have some rights as human beings."

Demographic change in the Hudson Valley mirrors the nationwide movement of Latinos to small rural towns.[58] Typically, the immigrant's path to incorporation, indeed Americanization, is through economic success.[59] Yet the rural job market offers only low-wage jobs, which are staffed by poorly educated, undocumented immigrants with poor English-language skills.[60] In other words, rural employment is not offering the sort of opportunities that have typically led to the social mobility of immigrants.[61] Moreover, the undocumented tend to earn less than other workers, and one study found that they were more likely to be paid below the minimum wage, twice as likely to be cheated out of their wages, and that wage theft on the part of employers was more acute for them than for other workers.[62]

As among any population of the undocumented, fear of the immigration authorities was by far the most common concern. It pervaded every aspect of the life of one fruit packer, who commented on the depressing omnipresence of this kind of anxiety. "It affects me in every way. I don't feel at liberty to drive. I can't go everywhere I would like. I am always afraid of being found by immigration." Workers like him usually kept a very low profile, which often meant not going out at nighttime, when their activities might have been perceived as suspicious by prejudiced law enforcers and locals.[63] Not everyone, however, was afraid of deportation. For some it would result in a long-needed visit home, while others insisted they would simply return to the United States after a few weeks. Many of my interviewees had friends and family who had been deported, and a few had undergone the rather abrupt experience themselves. "They just grab you and send you home," one said. Yet most of those I spoke with were quite nervous about being caught. Media coverage amplified the fear with sensational depictions of workplace raids.

One worker asked, "Haven't you seen on TV how they come to your door and then just take you away?" Furthermore, I heard widely circulated stories about immigration officials stationed at a Walmart, paying unannounced visits to labor camps, or creating checkpoints on roads. One worker described how his nephew was deported after he ran a red light: "He wasn't drunk or anything; he was just coming from work." Many also told me they were often stopped by the police and asked for legal documents. As one worker put it, "If they want, they can do anything with us; they really don't like us."

Aside from the fear of being randomly accosted by immigration officials, many were also concerned about being turned in by a neighbor, their employer, or a coworker. Even when workers thought employers would not dream of calling the immigration authorities, they clearly believed that their legal status greatly influenced the work relationship. Consequently, employee protest was rare, and walkouts, strikes, or collective complaints almost unheard-of. I found plenty of support for the perception that fear of deportation directly influenced behavior on the job. For example, fifty-one-year-old Alejandro from Mexico, who had been working in the East Coast migrant stream for three years, told me, "We are treated like unknown people but are not fugitives. We come here to do farm work because we do not have jobs at home. . . . We are not paid well and cannot ask for more." I heard from a worker on another farm about how their fear drives them to accept the work conditions offered. "They treat us like nothing; they only want the work. . . . Whether we like it or not, we have to like it." Even when they acknowledge this state of fear, many public observers consider the migrants' plight to be just another chapter in the hard-knock story of immigration and point to the remittances sent home, residences built, children's educations paid for, and other projects that the undocumented manage to fund with their wages. Javier eloquently summed up the predicament that many others had described to me: "When I am alone, I think, 'Even if the cage is golden, we are still locked up.'"

Food writers are fond of the notion that local farms play a role in creating community,[64] but that communitarianism clearly does not extend to the laborers. Because the undocumented prefer not to be noticed, they inhabit an underworld concealed from local communities and cut off from any avenue for accumulating social capital.[65] A worker I tutored at an ESL program in Dutchess County shared with me that although he had been living in the center of town for eight months, I was the first white person to speak with him. Farmworker families are also affected. Inez, an apple packer, lamented that her teenage children were not allowed to go out with their friends. "We're very afraid about the police getting them. My daughter wants to go to parties, the movies, the mall, but we don't let her. It's hard because they want to have fun, but it's better not to go out, for the fear that . . . " At this point Inez shook her head and grimaced, thinking about the dreadful consequences.

These farmworkers, many of whom were recent arrivals in the United States, evaluated their economic and social reality in comparison to their original homes and not in relation to other U.S. workers. This helped them rationalize their poverty-level wages and working conditions, which, they were well aware, would not be accepted by U.S. citizens. A day's pay at home was often less than the U.S. minimum hourly wage. Mostly poor and with little formal education, they hailed from rural areas in Mexico and Central America, where a lack of resources was the norm. "I used to have my own potato farm, but [now] there is no water. Nothing happens with land that is dead," reported an apple picker. Many explained to me that when they lived in their home country, all of their wages went toward purchasing food. Income from their U.S. jobs provided relatively well for families left at home. Many workers had built houses, dug wells, installed electricity and plumbing, and contributed to community-building projects from their wages. As a result, they were largely appreciative of the opportunity to earn poverty-level wages here that purchased a life of comfort there. This was reinforced constantly since they were very likely to work alongside others from their hometowns.[66]

Economic comparisons to their home countries make little sense for those who live in the Hudson Valley year-round. There is an increasing trend toward immigrant farmworkers settling in New York State.[67] My interview data attests to these longer sojourns in the Hudson Valley, in spite of many workers' initial intention to return home after a few years. When asked about their five- and ten-year plans, 75 percent of my interviewees said they planned to return to their home countries permanently.[68] Almost half of the workers had wives in their home country, and a little more than half had children there. Only 7.2 percent expected to continue working for the same employer. Six years later, however, I identified almost half of workers on the same farms and heard about another dozen who had gone home and were planning on returning to the farms the following year. None of the farmworkers I reinterviewed six years later had plans to return home in the next two years. Asked whether he had imagined he would still be on the same farm six years after our initial interview, a man whose wife and four children lived in Mexico acknowledged, "I never imagined I was going to last this long. . . . Now all my children are studying at home, and it's a lot of money. I had no other choice than to stay here. Earning in dollars here is better for us than earning in pesos there. So there you go."

Anyone familiar with the aspirations of newly arrived undocumented workers know that almost all plan to spend only a few years—*un par de años*—in the United States but end up staying much longer. There are many reasons for these prolonged stays: some get hooked on footing just one more bill, such as for medical expenses, schooling, or home improvement projects; others plan to save a larger nest egg; and some cite the improved opportunities for their U.S.-based children or pressure from their kids as a reason to stay. For example, a vegetable worker told me he planned to return home "soon." When I reminded him that he was repeating what he had said to me six years earlier, he laughed, "Of course! You always have to have this idea in your mind of returning soon." His plan had been to save money to build a home, but nine years and two children after he first arrived in the United States, he had not saved a penny; the portion of his pay that he had sent to his parents before he got married was now redirected to new household expenses in the United States. "I currently have nothing," he lamented. "Here you spend dollars, not like at home, where the money goes further. . . . You honestly cannot save money here. We always say we are going to return, but we don't because of the kids or another reason."

There is very little opportunity for advancement in farm work, and the workers I spoke to identified many obstacles that stood in their way of securing a better job.[69] Among those singled out were the absence of opportunities and a lack of job skills. Low levels of education and literacy constrained their search for better jobs and prevented them from acquiring jobs in service sectors such as retail.[70] The lack of English-language skills severely inhibited workers' ability to find jobs and advance in them. It prevented them from communicating effectively with colleagues, managers, and employers, and it generally prevented them from assimilating into communities outside their kin networks. One worker summed up what many explained to me: "No skills, no education. I can't read well; I can't do better." Many of these obstacles were related to workers' poverty and lack of resources for learning skills at home, as well as their reluctance to invest what little free time they had in acquiring these skills while in the United States.

Today's farmworkers are often regarded as participants in just another immigrant story, a tale characterized by family separation, hard work for low pay, and sacrifice, all in the name of improving the prospects of the next generation. Although some of these themes rang true for the workers I interviewed, comparing them with immigrants of earlier periods is misleading. Opportunities to earn citizenship and to climb the job ladder are much more tightly circumscribed today than they were in the nineteenth and early twentieth centuries. The subjects of my study had little opportunity for job advancement or to earn a green card or citizenship, even though so many were settling permanently in the United States. Perhaps more important is that although the individual workers might return home or find better work in construction, landscaping, restaurants, or domestic jobs, their farm jobs would quickly be taken by new immigrants, creating a cycle that would further institutionalize the vulnerability and poverty of the state's farm work force. This is not just another immigrant story; this is a story about reproducing inequality among the workers whose labor is fundamental to food provision. Any code of ethical eating that ignores this perpetuation of injustice is highly selective, if not morally hollow. A comprehensive food ethic would privilege individual human workers as much as animals and the environment.

THREE

The Farmers

CHALLENGES OF THE SMALL BUSINESS

TO FULLY UNDERSTAND THE STRUCTURAL plight of Hudson Valley farmworkers, we must take a step back and explore the pressures and predicaments facing the region's farmers. Why do farmers treat their workers the way they do? This is by no means a simple question to answer. Because they are confronted with constant economic challenges and are forced to balance an array of obligations, farmers' own positions cannot be lightly dismissed. Insecurity has been a condition endemic to Hudson Valley farmers since the time of tenant farming. In the regional record, it is difficult to locate an era of more than a decade or so of thriving, profitable agriculture; nor is this instability unique to the region. My interviewees were constantly on the lookout for new marketing avenues and alert to changes necessary to do business. This mentality was sometimes presented as nerve-racking; not a few told me they were worried sick at the prospect of not surviving the year. Certainly this is a perennial anxiety for most small entrepreneurs.

Why should farming be an exception? Is the agricultural business appreciably different from other small businesses? Clearly, the investment required can be daunting. Even when one "starts small" there are expenses for the purchase or lease of land and equipment and considerable outlays for technologies of soil improvement. One must also consider that many farmers live and work on the same acreage. Unlike founders of high-tech startups, farmers usually do not have angel mentors to see them through the fledgling years. Rather, insecurity has been a constitutive feature of farming life for so long that farms are expected to fail. These ideas were described to me as part of the peril and pride of growing food and were specifically linked to the agrarian ideal, the resiliency of farmers, and the unique status of the profession that demarcates the farmer from other businesspeople.

Relying on farmers' testimony, this chapter aims to present a deeper, empirical portrait of their role in the regional food system, from the daily challenges of running a farm to the opportunities offered by the local food movement. In doing so, I will try to describe their economic predicament, further examine the ideological resources that help them construct a portrait of the industry, and, last but not least, shed light on their conflicted relationships with employees. The fresh scrutiny generated by farm-to-table consciousness has heightened public affection for the idea of farming. Examining local food opportunities also intersects with some of farmers' most pressing concerns, such as the threat of foreign competition, the encroachment of real estate development, and the capriciousness of community support. I also analyze farmers' opinions on labor management to offer a more nuanced portrait of their self-image and functional role as employers. In this chapter, then, labor concerns are presented from the perspective of the employers who experiences chronic difficulty not only in securing reliable staff, but also in supporting their mostly seasonal work force.

PORTRAIT OF HUDSON VALLEY FARMING

Although operational practices differ according to the size of the farm, whether the farmer engages in organic or traditional pesticide practices, and the types of plants tended, farming is always influenced by regional factors. These include the availability and quality of land and water, local transportation, types and sizes of markets, land development patterns, and the prevailing labor market.[1] Though I interviewed a diverse range of farmers, I focused on those most likely to hire seasonal workers, who have historically been the most vulnerable work force. I consequently steered away from farms that primarily raised animals, although one of my interviewees sold chicken eggs, another sold chickens and cheese, and one was experimenting with raising pigs.

The vast majority of my interviewees grew fruits and vegetables, and some were involved in nursery production. Those engaged in organic practices were all new to farming.[2] More than half of the farmers had their farms handed down through the family or their spouse's family. Farmers varied in age from their late twenties to late eighties, a minority were women, and almost all appeared to be white. While farm size varied from eleven acres to several hundred, some growers also leased additional land. They sold to a variety of markets including CSAs (Community Supported Agriculture), farm stands, farmers' markets, restaurants, regional wholesale markets, and large retailers such as Walmart.

All but one of the farmers I interviewed was a family farmer, and only one ran what could be called a small farm. While the terms "small farm" and "family farm" are often conflated, they mean quite different things. Definitions of farm size are not related to the amount of land farmed. Dairy farms in the Hudson Valley were typically large to accommodate grazing. Today's orchards and farms raising baby vegetables, however, do not take up nearly as much room as pasture fields. The USDA defines small farms as those with gross sales below $50,000, medium-sized farms earn $50,000 to $249,999, and large farms gross $250,000 and above. For the six Hudson Valley counties I studied, 72 percent of farms earned less than $25,000 in sales, making them very small. Of the remaining 28 percent, more than half earned above $100,000 in sales a year.[3] Most of the farmers I interviewed ran medium-sized farms and three had large farms. In other words, the farmers I interviewed were not tiny mom-and-pop operations, and all used hired labor.

The term "family farm" refers to two main characteristics. First, a family farm is owned by an individual or a family as opposed to a nonfamily corporation. Second, it is run by the family and not a hired manager.[4] The term does not imply that a farm was handed down through the generations. A new farm can be a family farm. That said, my interviewees attached different values and notions to the concept of family farming. One vegetable grower described how the expansion of his family farm operation had turned him into a manager, or, as his sister put it, a CEO, whereas two decades earlier all the work had been done by his father and himself. Another emphasized the intimacy of the operation. At least once a week, he explained, one of a handful of older local men, most of whom had grown up on farms themselves, dropped by to inquire about how things were going. During another interview I sat in the living room of an old farmhouse with the air-conditioning set on high for the benefit of sacks and crates of potatoes, tomatoes, corn, cucumbers, and zucchini. The farmer explained that this was more cost-effective than turning on the refrigeration in the barn. He expressed his doubts about the future of his operation and described family farms as a "dying breed." Many who spoke to me about the quality of their work life cited pride in the family legacy of their trade to explain their choice of livelihood, saying, "It's in my blood. I grew up with it." Others spoke of their love of the outdoors and the virtues of being their own boss. One young fourth-generation farmer mused about the romance of intergenerational farming,

"It sounds cheesy, but having a connection to a place, and feeling like that particular place is important and meaningful, is very real to me." This feeling, he concluded, is not easy for most people to understand since so few Americans live where they were born or have any sort of ancestral connection to their homes. He continued, "You can feel it if you go somewhere where a family has a cabin that their grandfather built, for example. That connection is palpable; you can feel that there is something there."

Local Farms: Thriving or Surviving?

Rising land prices, dwindling community support, competition from overseas, and the high subsidies paid to the country's largest farms all have made it difficult for individual growers to squeeze out a good living. Government and nonprofit reports on the status of Hudson Valley agriculture document the ongoing loss of farmland, increasing foreclosures, and the shift from arable crop production toward horse rearing, greenhouses, and landscaping and gardening operations.[5] Under such conditions, and in the face of so many challenges, the regional farmers I met with attested to the struggle to consistently make money. "You have to eat, drink, and sleep the bottom line—profitability," concluded Bobby Trask, a Hudson Valley farmer with an extensive and thriving fruit and vegetable business.

Often, being business-minded was at odds with the passion many farmers had for their vocation. One farmer explained to me, "Unless you've got that fever in your belly to make a buck, you don't survive." He concluded that his biggest mistake as a second-generation farmer was seeing himself as the "caretaker of the family garden," engaged in cleaning up a hedgerow or planting trees along the street. But taking pride in the upkeep of the landscape was at odds with breaking even. "With the international cheap food policy," he observed, "there is no room for that. You have to chase the dollar." According to the well-documented history of Hudson Valley farming, this grower was many generations away from the era of subsistence farming, when agriculture was not driven by profit. Sentiments such as his typify the perception that each generation has it harder than the one before. In this case, he compared his situation to the days when it was easier to survive by selling wholesale. Even as late as the 1950s, commissioned salespeople from private trucking companies would arrive on the farm to bid for fruits and vegetables.

No such opportunities exist in the twenty-first century, and surviving on wholesale transactions alone is next to impossible. Some of this is related to the rising costs of equipment and more recently, I was told, to a threefold increase in the cost of operational inputs such as fuel and fertilizer. One veteran vegetable farmer recalled that he used to trade in older farm machinery for the latest model but that he had given up the practice because the new machines were too expensive. Even unexpected windfalls could have adverse consequences. In 2007 and 2008, the apple crop in China was off, and as a result cider mills were paying twice as much as usual for apples. The resulting higher prices for cider and apple juice, however, meant that consumer interest waned, and the market was oversupplied. "You never know what's going to be the thing that makes you go out of business" was a recurring theme in my interviews. Although it is never just one factor that determines the fate of smaller-scale farms, in my decade of research in the field I often heard about how the latest opportunity would "save farming" or the most recent obstacle would "kill farming."

Even the new lease on life offered by farmers' markets and local consumption had the side effect of increasing competition and friction between farm owners. Opening a staffed farm stand in a key location, I was told, could threaten the profits of nearby farms. In one town it was rumored that the most successful local farmer was trying to block the establishment of a nearby farmers' market, which would eat into his profits.

Strategies for improving the Hudson Valley's farm economy included creating a distinct regional food identity, incorporating value-added products to the farmer's repertoire, developing direct and niche marketing, and fostering regional cooperation through new business strategies. The emergence of new farmers, or "greenhorns," could be counted as evidence of a regional renaissance in farming. Most of these greenhorns started with very small farms or leased land because they could not afford to buy large parcels. Some were involved in producing artisanal or boutique products, while others, particularly those who specialized in baby vegetables or greens, had opted for high-density farming. One of this new breed told me how he grew greens for local chefs, a relatively new customer base that did not purchase directly from farmers with any regularity until the 1990s. Without restaurant purchases, he informed me, boutique organic farming in the Northeast could not support a family. In addition, swelling membership in CSAs was providing a stable source of support for some new entrepreneurs.

As I witnessed these changes starting in the early 2000s, I also saw how traditional farmers were diversifying their operations with more types of produce and value-added goods and shifting away from wholesale toward direct sales to consumers. Transitioning toward more diversity is one of the latent strengths of local systems. Writing in an earlier era about changes in agriculture, one observer presaged such a practice, pointing out that catering to a local market required diversification,[6] while another found crop variety to be a hallmark of the family farm.[7] The local sales implications were clear; Bobby Trask, for example, explained how he increased his leverage by producing many varieties of fruits and vegetables, and he claimed that he could "put on a really nice show of Hudson Valley produce" at farmers' markets.

Valorizing Farming

Food writers are known to exalt the romance of farming, but farmers themselves are also quite adept at polishing a pastoral version of the American dream. Walking through a defunct dairy barn that he uses for storage, Bobby Trask took a deep breath before saying, "I still sense the sweet smells of this barn—the open area, the milking parlor. It brings back a lot of memories." Trask's childhood remembrances of the farm evoke the idyllic rural life we all recognize from novels, advertising, and films: "On my first day up here, I was with a bunch of farm kids. They were taking off their shoes and socks and saying, 'Come on.' I did, and we walked through the barn and felt the manure through our toes." He pointed to the hayloft he played in, showed me the cornfield he navigated to get to school—using the telephone lines as a guide—and described trying to ride six-hundred-pound calves. "We used to stand on that fence, jump on their backs, and see how long we could hold on." During this tour, he discussed farming techniques with me as he checked on his crops and workers. Much of the topography of his rural Hudson Valley town had observable sentimental value for him, as it conjured up the halcyon freedom of his barefoot youth. Those childhood experiences left him with a strong attachment to the land and respect for its potential. Looking across the fields to a housing development, he lamented the loss of fertile land. The soil, he said, was among the best in the region, continuing, "If I had enough money I would buy those houses and knock 'em down."

Trask's zealous dedication to farming life was also suffused with the belief that his livelihood was one of substance, satisfaction, and the kind of accomplishment that is missing, he says, from jobs that involve pushing paper, tapping keyboards, or making deals. He conceded that construction or working on cars might be worth pursuing as occupations but claimed that they did not generate the sort of gratification that came from working the soil. The veneration of this kind of toil is the legacy of farming in the United States, and humble, even poverty-stricken farmers have always been considered worthy of esteem, while those born into wealth have been seen as lacking access to the benefits of the rural work ethic.[8] Versions of this conviction about the virtues of skilled hands-on labor cropped up again and again during my interviews.

Much of this sort of testimony was also imbued with the moralism that is one of the main tenets of agrarianism.[9] Consider the following examples from *The Greenhorns,* a film by a Hudson Valley farmer-visionary about "the redemptive force" that young alternative farmers offer "for society, culture, and agriculture." In the film, one of the new wave of farmers remarks, "I'm not relying on some strange economic structures that have been set up that benefit some and hurt others to make my livelihood." Another declares, "I could actually do something that I'm morally down with and still pay the mortgage."[10] Such statements appeal to agrarian codes of wholesomeness and integrity. For the greenhorns especially, the lack of big profits allows them an imagined distance from capitalism. In other words, the land can be worked for commodity produce, but as long as it generates a meager living, the practitioners can feel that they are uncorrupted by the market.

This artisanal zeal for small farming is shot through with agrarian nostalgia for self-sufficiency and a less complicated life. To some degree, the local food movement is crafted around the resurgence of these ideas.[11] Quite regularly now, Sunday newspaper supplements carry real estate articles about the desire to trade in the grind of urban living for a small farm where one can slow down, enjoy "the simple life," and live more authentically.[12] The trend has also spawned several books on the trials and joys of relocating to rural destinations.[13] These books, which range from the humorous to the pretentious, divulge personal transformations related to connecting to the land and the life cycle.[14]

On Hudson Valley fruit and vegetable farms, where the produce must be carefully harvested by hand, it is possible to imagine a recherché livelihood that is liberated from modern convention, but that would be to overlook today's sophisticated operations that depend upon research and technology so that farmers can make decisions about plant breeds, pesticide application, and planting schedules. One farm I visited, for example, had multiple weather stations to provide information for such decisions. Newcomers should also be warned about the low compensation offered by farming. Not a few of my interviewees compared their vocation to a crapshoot or a trip to Atlantic

City. One veteran declared, "I don't gamble with lotto or anything like that. We gamble on the farm, and with luck, we win. This year with the weather as it is, we lose." He turned to his son-in-law, who nodded in agreement and added, "Two steps forward, three steps back," and they both laughed. Such sentiments are encapsulated in the common aphorism that farmers live poor and die rich. Indeed, researchers examining farm household wealth as opposed to income show that 94 percent of farm households are better off than the average U.S. household.[15] Typical farmers, however, have invested so heavily in their farms that their day-to-day fear about crushing financial burdens is also palpable. A joke commonly told in the area is about a farmer who wins the New York Lottery jackpot and, when asked what he'll do with the winnings, responds, "I'll keep farming 'til it's gone."

THE PROMISE AND CHALLENGE OF LOCAL FOOD

"FORGET ORGANIC. EAT LOCAL," screamed a 2007 *Time* magazine cover story that summarized the insider consensus that local was increasingly the choice of "food purists." This story and many others like it ushered consumers away from purchasing corporate or foreign organic produce in favor of buying from local farmers.[16] Undoubtedly, the interest in local food stems from what scholar Melanie DuPuis describes as one of the main ethical consumption questions addressed by food writers: "Can we eat our way to a better world?"[17] Consumers are interested in a range of implications of their food choices. Much was made of the revelation that the Whole Foods colossus purchased organic produce from China—where U.S. eyes have long been trained to cast suspicion[18]—and that U.S.-grown organic bagged spinach was recalled after a 2006 *E. coli* outbreak that killed three and sickened two hundred.[19] Worries also grew about the carbon footprint created as a result of the mileage that California and foreign organic produce traveled en route to discerning palates elsewhere.[20] In addition, as organic products became more popular, concern mounted about the consolidation and corporate cooptation of the industry sectors that supplied them.[21]

The interest in local food, pitched as an alternative for those dissatisfied with food from corporate agriculture, has brought increased attention to regional farmers. The farmers I interviewed were eager to discuss the differences between their products and their well-traveled competitors, and they had a fierce distrust of other countries' agricultural practices. Growers told me that I might be consuming three-year-old irradiated peppers from Latin America,[22] and that farm chemicals banned in the United States found their way from the U.S. government to the developing world and back to my dinner table. One farmer pointed out that international organics were not directly inspected by the USDA. Instead, he claimed, "The task is outsourced to corrupt third-party certifiers."

Of all the changes to Hudson Valley farming in the last few decades, the "buy local" movement seems the most promising for sustaining regional farms. As a *New York Times* travel article declared, "Food has been the great engine in the region's revival," and the Hudson Valley "has now gone thoroughly locavore."[23] Amtrak's onboard magazine *Arrive* (which I picked up on the Hudson River line) proclaimed the Hudson Valley a "culinary destination."[24] These are examples of the promotional apparatus that helped give birth to the Hudson Valley's distinct food culture, which is rooted in local food sales, artisanal value-added products, and the pastoral rural vernacular of the landscape that connects the era of yeomanry to today's greenmarket-focused growers.

Much of the praise for local food is linked to nostalgia for self-sufficiency, a lack of commercialization, and the innocence of rural life.[25] This is typified in an article in the magazine *Edible Hudson Valley,* which claims, "Traditional agricultural values of generosity and cooperation, rooted in the Hudson Valley, seem infused in each batch of cheese produced."[26] The main vehicle for reinforcing these agrarian sentiments and promoting Hudson Valley food culture is the high visibility of farmers' markets and farm stands. Scattered throughout the region's small communities, they not only advertise the farms' wares but also offer visual reminders of the scale and diversity of the industry.

In the past decade, farmers' markets have spread to multiple urban locations and provincial towns in and near the Hudson Valley.[27] Local sales to consumers, restaurants, and local distributors result in more profits for the producers since the middleman is cut out.[28] One Dutchess County farmer I interviewed participated in as many as forty markets a week and fielded invitations from more than he could manage to attend. Many of his peers also had farm stands with clearly marked prices and a lockbox for customers' cash. Others ran more sophisticated staffed operations, often near downtown areas or in other strategic locations, that sold value-added goods such as cheeses, maple syrup, honey, and pies.

Much is made of the role of individual farmers who sell locally. Although the agrarian ideal draws on the memory of the independent owner who does not rely on the market or industry networks,[29] today's circumstances offer a twist on this concept. The local farmer may be able to operate outside the industrial food system by making independent decisions about what to grow and where to sell, but she is still tied to a market—in this case, a local one. Yet this local market allows customers to have a relationship with a farmer that is impossible for shoppers at a grocery store. And, if the customer wants questions answered, direct access to the farmer is more or less guaranteed. For example, farmer Bobby Trask described to me his visit to a twenty-thousand-acre vegetable farm in California, saying, "The guy that gave us the tour couldn't answer two-thirds of my questions." In contrast, Trask emphasized that he could answer 95 percent of questions about his own farm. And questions, he added, were all too often asked by consumers, who took pride in their knowledge about regional produce. Another farmer who ran a CSA and pick-your-own operation brought it back to the ethics of the face-to-face relationship, which he believed made an appreciable difference compared with the corporate industrial farmer: "The people who sell direct to the customer have to justify each of their agricultural practices. When you have that scrutiny, you clean up."[30]

Food writers concur. One of their recurring themes about the individual farmer is that the consumer gets, as Michael Pollan reports, the "guarantee of integrity" obtained "when buyers and sellers can look one another in the eye."[31] This kind of sentiment and self-marketing is aimed at positioning the local farmer not as a businessperson, but as a community peer. Indeed, one Hudson Valley farmers' market customer told a reporter, "When I get the additional information I'm seeking from a farmer who is my neighbor—not just a commercial seller—it makes the food I buy that much more special."[32] So how often does the consumer get to look the farmer in the eye? Local farms might bring to mind heartwarming images of mom-and-pop operations, but the reality is that many farmers hire others to run their stands at farmers' markets. At New York City markets especially, those looking you in the eye might not ever work on the farm. Even at local farm stands it is rare to see the owner, though it is customary for a family member to be present.

Moreover, it is the larger farms that dominate the local sales economy. If we define local sales as those that are direct to the consumer, then on a national level, small farms account for only about a third of sales. When the definition of "local" is expanded to include intermediary sales to restaurants, grocers, and regional distributors, large farms are dominant: in 2008, about 87,000 small farms (profits below $50,000) earned 11 percent of the total of local sales, while 15,202 medium-scale farms (profits between $50,000 and $249,999) accounted for 19 percent, and 5,301 large farms (profits $250,000 and above) were responsible for 70 percent.[33] This confirms what other authors have estimated:[34] that direct-to-consumer sales make up a small portion of the local food economy[35] and are much less significant than food writers would have us believe.

Competing with Cheap Food

Trask took the time to explain the important concept of the "loss leader." This is a marketing strategy that offers a desirable product at below market value with the goal of getting customers in the door so they will spend on more profitable items. When Walmart was selling blueberries at 99 cents a pint, Trask joked that he should be buying his blueberries there and reselling them at his farm stand. In more sober tones, he reiterated what he told his customers: he could not compete on price, but he was offering what Walmart could not—"quality, variety, and produce that was picked the day before." He also affirmed that local food dollars support not only a local farmer, but also the preservation of open space, which is of benefit to the community at large.

Although local foods are attracting a growing portion of the consumer market, for the most part Americans are habituated to the lower cost and year-round availability of produce they purchase at the supermarket. Indeed, residents of New York City and the greater Hudson Valley spend only slightly more than 1 percent of their food dollars on regional produce. Yet farmers and marketing experts see tremendous opportunity for growth. Census data showed a 36 percent increase in direct-to-consumer sales within the greater Hudson Valley between 2002 and 2007. One report appraised consumers' spending power and posited that if regional consumers spent 10 percent of their grocery budget on local goods, Hudson Valley farmers could bring in $4.5 billion in sales (nine times what they were earning in 2007).[36] On the other hand, 1 percent seems almost insignificant. There is a consensus that "buying local" is primarily an upper-middle-class trend adopted primarily to the cognoscenti informed by Hudson Valley–based apostles of locavore eating, such as food studies' godmother Joan Dye Gussow, and Dan Barber, renowned chef at restaurants in both Manhattan and the Hudson Valley. While farmers markets often accept food stamps, they can never compete with the prices of large supermarkets or with the convenience of shopping there.

Even within farm households, the switch to eating local can be a challenge. During an interview at a farmer's kitchen table one late July, I asked about the origins of the fruit in a large bowl that was in front of us. "All over," he replied as he picked out pieces of fruit labeled "New England grown." "Bananas? Sixty percent are from Ecuador." He read the label, "Dole. Yes that's a favorite of Ecuador. Grapes at this time of year? They could come from Southern California." Even though he was resolute in his own commitment to eat from his own fields as much as possible, this bowl was for his teenage son, who had lobbied for more diversity on the table.

Challenges considered more serious than drawing in customers were foreign competition and the nation's cheap food policy. In this regard, many of my interviewees focused on the health dangers posed by foreign produce. And virtually every grower I spoke to asserted that consumers' expectation of cheap food was a great disservice to farmers. Trask reported that at a Manhattan farmers' market, "People drop $20 to $30 on flowers and then they nickel-and-dime me on tomatoes." Growers also blamed the U.S. government for fostering cheap imports and for subsidizing huge factory farms at home to keep prices down. "Look at any government," remarked an orchard owner who was convinced that politicians were primarily concerned with self-protection. "If there is a food shortage or if the price of food spikes, it is proven in history that the people will unseat the politicians."

Many of the farmers I interviewed explained that a greater percentage of Hudson Valley produce was sold wholesale before retailers found cheaper produce from abroad. Apples are a case in point. New York is the second most important apple-growing state, after Washington, and apple orchards have been a staple component of the Hudson Valley landscape for more than a century. These farmers, however, must now compete globally. Until the late twentieth century, New York apple growers had a ready local market for fresh fruit, while large brands like Mott's contracted with them for fruit for applesauce. Even bruised and fallen fruit found a home on the apple juice concentrate market. In the late 1980s, however, neoliberal trade policies drastically altered this stable transactional market, allowing apples first from New Zealand, and later from China, to flood the domestic market. By the mid-1990s, New York had more or less lost the apple concentrate market to these international competitors, and the region's farmers are still mourning the loss today. Foreign competition arrived just when many orchard keepers were testing new types of apples, including old-fashioned varieties. If they were no longer going to succeed based on price or bulk production, these growers would now offer specialty products, as they had in the early 1900s. Once again, the region's farmers were forced to change their tactics to stay in business.

Community Support?

A benefit of the local food movement that is prominently discussed is the role it plays in fostering community. We are told that farmers' markets and CSAs promote socially networked activities and a sense of community; that local agriculture stimulates trust, reciprocity, and invigorated social lives; and that, unlike food tied to the global capitalist food system, local food is a form of "civic agriculture" that keeps profits within the community and encourages more active community engagement. Michael Pollan called farmers' markets "the country's liveliest new public square."[37] Another author brings us right back to the Jeffersonian ideal by arguing that local food systems can be a vital tool for promoting social equity and democracy.[38] How did my interviewees regard the community benefits they offered, and did they consider this to be an inherent obligation of their livelihood?

The region's farms and open space are often touted as an antidote to suburban or urban living, yet I discovered that many newcomers to rural life did not understand or fully appreciate what it meant to have farmers as neighbors.[39] As the threat to farming from development increased, so did legislation to protect farms. For example, New York's state legislature passed the Agricultural Districts Law (1971) to create special agricultural districts with the goal of promoting farming. This law provided for agricultural assessment of the "use value" of farmland (as opposed to market value) for property taxes as well as required that new government policies (local and state) support the viability of farming.[40] New York also became an official "right to farm" state with the Agricultural Protections Act of 1992,[41] with further legislation enforcing that status in 1998.[42] What new neighbors rarely know is that this type of legislation discourages nuisance lawsuits against farmers for issues such as manure odors, machinery noise, and pesticide application so long as the farmer engages in "sound agricultural practices." Other laws help farmers by offering them school tax credits as well as tax exemptions on farm buildings, farmworker housing, and sales tax.

Despite the increased emphasis on how local foods promote a sense of community and bring farmers and neighbors together, my interviewees expressed no end of concern that the future of farming was compromised by a *lack* of community support, particularly from new homeowners. One fruit grower in Dutchess County argued that the right to farm laws only go so far and that strong community allies were really needed. He said, "They are enamored of the agricultural land next door, yet they have lots of questions, and we're the people who answer to them." When homeowners drop money on their exurban spreads, they tend to feel entitled. A fruit grower paraphrased a typical complaint: "I spent $500,000 on my house; we don't want all that spraying near us!" Even the spread of lime, I was told, could inspire horror: "Oh my god! What is that pesticide you are spraying here?" He reported that his seventy-six nonfarm neighbors were mostly supportive, but if each had a question about his operations once a year, that added up to a call every five days. He offered some context for this low-intensity conflict by explaining that in other states, like Iowa, neighbors are better informed about farming, so if you have a dry year they might sympathize. In the Hudson Valley, however, he told me, "You are running pumps twenty-four hours a day to make something happen, and your neighbors are saying, 'When are you going to turn that thing off? I can't sleep at night.'" Trask, who had largely moved to the application of pesticides by hand, told me he still liked to use a crop duster occasionally, "just to make sure people don't forget where they live." He was not joking; he wanted to remind his neighbors that he has the right to spray from the air, especially since neighbors in new houses were increasingly complaining. Like most of those I interviewed, he blamed realtors for not fully disclosing to prospective buyers the rights of farmers and the drawbacks of buying land near agricultural operations. Lack of understanding came in all forms. One farmer who had leased adjoining land for a couple of decades put it this way: "The new owner is an artist. She doesn't want me to remove the old fruit trees because she likes to draw them. That doesn't make it easy for us."

THE INTRACTABILITY OF LABOR

Of all the economic challenges that make farmers' lives difficult, making payroll is the most contentious. From a purely budgetary standpoint, labor is an expense (often the largest line item in spending), and workers can easily be viewed as "labor inputs" or "man-hours." Yet farmers are often involved in the lives of their workers as more than just employers; they are also landlords, transportation providers, and protective bulwarks against hostile community members. The topic of labor was so sensitive that several farmers hung up on me, and a few notably raised their voices when I mentioned my intent to interview them about their relationships with their employees. Although some offered to refer me to other farmers, one warned that the farm owners he knew "would tell me where to put my microphone." When farmers were willing to discuss their labor practices and problems, they revealed their concerns and opinions on some of the major labor management issues they face. The most salient challenges included establishing wage levels, the cost of maintaining employee housing, and the exhausting task of dealing with government labor regulations. In addition, finding and maintaining a reliable work force was a perpetual concern. New immigrants and the undocumented, although potentially good workers, also brought with them a host of new issues that had to be handled.

I spent a workday with Trask, who offered to take me along with him on the condition that I arrive early. While it was still dark, I made my way to his farm, where he was warming up an old truck for a trip to the dump. He was the only person around at that hour, but when we returned from unloading the refuse, workers were preparing for the day and loading empty bins into trucks. The bins would return at the day's end overflowing with vegetables. The atmosphere was congenial and the workers joked with the boss. As the day went on, the *norteño* music in the building would grow louder. One young worker sported a dangling silver earring. Trask did a double take, smiled, and asked, "What the hell is that in your ear?" He went on, "Can I allow that? Is this a safety concern?" The others laughed. Trask, who liked to joke, kept up the banter for a few more minutes while reviewing the picking slips with his lead workers. He felt strongly that computers were an invaluable tool for the modern farmer, but some aspects of his business significantly predated them. Harvesting was still guided by the picking slip, a two-by-five-inch photocopied list with tiny handwritten abbreviations for the crops next to the quantity, added by hand by Trask each morning. The template had only been revised three or four times in the past twenty-five years. Throughout the day Trask checked in with worker-supervisors, and he followed up with other workers about airfare prices he had found for their trips home.

The good humor, Trask explained, lasted as long as the work was going well. If there was a problem, his temper could flare, and he was quick to express anger. After hiring some teens a few summers before, he recalled, "I kind of made a joke that some of these parents should sign a waiver that they won't sue me for verbally abusing their kids. But at the end of the summer, I got a bottle of wine and a card from one of the mothers thanking me, because she saw a big improvement in her son's attitude." Managing his farm had gotten more complicated as the operation grew. Trask knew what work was underway each day, but when something went wrong, it could be difficult to diagnose the cause. His workers and supervisors protected each other and were reluctant to point fingers. Trask recalled that after a forklift mishap, "I got so mad, I lost my voice. Nobody would say who did it. I would have gotten upset, but not as upset, if I had found out who did it. That's when I fly off the handle." The employees' reluctance to reveal the person responsible for mistakes seemed to be an expression of solidarity. Yet Trask also described how solidarity (not his term) was crucial for productivity since it fostered workplace teamwork and the ability of workers to keep each other on task with an attitude of "If I'm working hard, you better be working hard." Yet this kind of solidarity could pit employees against the boss, resulting in the employer feeling a loss of control.

Compensating Workers

My interviewees offered many reasons why agricultural wages were low and why they should be kept that way. One reason that was repeatedly brought up—both by those who sold to wholesale markets and those who sold direct to the consumer—was the very small profit that farmers made. The surplus value of agricultural products, they told me, was remarkably small. One interviewee acknowledged that both farmers and farmworkers were victims of the U.S. cheap food policy. He reminded me that, although Guatemalan workers were paid only about a dollar an hour, in traditional supermarkets, Central American produce sat alongside Hudson Valley–grown produce. Because U.S. food prices are so low, the growers were inclined to conclude that they had little to no flexibility for increasing wages, particularly in light of rising fuel costs and the increasing prices of other farm inputs, such as new machinery and pest-control measures.

Another recurrent theme was that agriculture could not easily be compared to other industries and that it had to be treated as a unique case. When quizzed about wages, Trask pointed to the idiosyncratic aspects of farming, saying, "Weather, disease, insects—sometimes it is a crapshoot. There are a lot of variables that affect our net. In construction they are charging the customers a lot more per hour so they can afford to pay higher wages." The comparison to the construction industry came up often in my conversations. Farming and construction are similar in that they are largely seasonal, they are huge economic sectors, and they hire significant numbers of undocumented immigrants, even though construction is often perceived as a unionized industry that hires native-born workers.[43] This difference in the construction industry's labor market was an important point of reference for at least one of my interviewees. He argued that exempting agriculture from overtime protections made sense because immigrants are not taking anyone else's job away, whereas in construction, immigrant workers might be undercutting native-born workers who would earn twenty to twenty-five dollars an hour. In other words, requiring overtime for construction work protected Americans, but in agriculture there was nobody who needed that sort of protection.

The wage rates farmers quoted to me varied, and not everyone was comfortable divulging the exact amount they paid. Several reported that they paid most of their workers in the range of $7.50 to $8 per hour but that a few longtimers and worker-supervisors might make up to $13 an hour. I also met organic and boutique farmers who paid their apprentices and workers more competitive wages. One, whose organic operation was thriving, paid around $12 an hour, and another, an artist-turned-farmer whose produce was sold mostly to high-end restaurants, hired college-age workers at a starting wage of $13 an hour for part-time work or up to forty hours a week, and as much as $16 an hour once they had proven their worth. "If they are good, I want to keep them," he reported. "I feel like I make enough money that I can pay that, and they are working hard. I think a lot of them do appreciate it; they feel it is good pay." He did not see himself competing for workers with other vegetable farms, and he reported that when immigrant workers applied for work with him, they sought housing and more than forty hours a week of work, neither of which he offered.

Most of the farmers I spoke to provided housing and pointed out that wages were only a portion of the labor costs incurred by employers. Housing on the farm was often free of charge to workers and usually included utilities. Trask indicated that heating costs alone for his farmworker housing ran him about $12,000 a year. Added to that were electric bills, trailer upkeep, and transportation, and some other farmers also provided telephones and cable or satellite television.

If farmer-provided housing was a perk of the job and helped employers recruit workers, it could turn into a liability for owners. After all, labor advocates and reporters often pointed to examples of decrepit housing as a visible sign of poor work conditions and a lack of respect for laborers. "They made it sound very *Grapes of Wrath*," complained one farmer about how a TV journalist described the housing he offered his employees. He paraphrased the journalist as saying, "'Only fifty-five miles from New York City and people live like this.'" He winced and explained that upkeep could be a challenge when workers did not keep their homes clean or broke household items. In turn, employers became less motivated to make repairs. A Western New York farmer, Joe Keen, spoke angrily to me on this topic and contended that he had little control over the workers and contractors who used his housing. A labor contractor he knew had shown up with a crew for another farm and moved into Keen's labor camp without his permission. Should he let them stay or throw them out? "I should just torch it," he barked about the housing.

New housing was expensive, I was told, and the replacement of run-down trailers or refurbishing of other homes would often be deferred due to a lack of funds, even when it was clear that upgrades were in order. Yet one particularly run-down and dirty house with an atrocious kitchen where I interviewed workers in 2002 had been significantly renovated by the time I returned in 2008. Although federal government loans, some at zero interest, were available to build new housing, the occupants needed to be citizens or hold a green card, and so, I was told by more than one farmer, the program was of little practical use. Farmworker housing was also a source of conflict with neighbors. According to an anecdote that I heard, a Columbia County farmer wanted to put in new housing with USDA-approved funds, but his neighbors organized and blocked the effort.

In general, housing regulations were regarded as a headache by employers, who had little desire to act as landlords. For example, workers often removed the batteries from smoke detectors in trailers because their cooking constantly set them off, but it was the employer who was liable and could get fined for the housing violation. One farmer, frustrated by the responsibilities he was facing, described to me in detail how worker housing had to adhere to very specific regulations. A house across the street from his current labor camp was going into foreclosure and would have served the needs of his staff, but the ratio of windows to floor space in the bedrooms was not sufficient to meet the legal requirements. Given these disincentives, it is no surprise that fewer and fewer growers were offering housing to their workers.[44] Trask described a different problem. Without asking him, several of his workers brought their wives to New York to join them. Although he considered this a natural desire, offering housing to wives and children was a financial strain and he felt he had to put a stop to the practice. I also knew of at least one case in which a worker was deported but his wife and children continued to live in the worker housing, putting the farmer in the awkward position of having to decide if they could continue to live there.

Housing was not the only aspect of farm labor that growers felt was over-regulated. Among those who conducted inspections on farms and played a role in monitoring labor practices were the county Department of Health, the federal Department of Labor Wage and Hour Division, and the state Department of Labor Rural Employment Division. In addition, each state had a monitor advocate within the labor department whose job it was to ensure that migrant and seasonal farmworkers received "qualitatively equivalent and quantitatively proportionate" work force development services. Such development services included job training for nonagricultural work, which some farmers saw as damaging to the industry because it helped workers leave agriculture. While county and state regulators were often perceived as helpful, federal regulators were viewed quite differently. In the words of an older couple who owned a fruit orchard, "The first time the feds came in, they wanted to see every piece of paper relating to workers' pay. They had guns in their holsters and we felt so vulnerable." Another common complaint was that the instructions for payroll paperwork were too complex to follow. More than one farmer also reported that hiring guest workers "puts you on the map of the federal Department of Labor," thereby assuring additional government oversight. In addition, some regulations, like the window-to-floor ratio required for worker housing, were seen as both excessive and needlessly expensive. Different interpretations of standards were also confusing. One farmer explained how he was required to have portable toilets with hand-washing facilities in the fields, but he heard that a farmer in the next county was required to have the faucets and soap *outside* the toilets.

Finding American Workers

Finding workers was universally cited by my interviewees as a tremendous challenge. For almost two centuries, farmers in the Hudson Valley have struggled to secure an appropriate work force for their seasonal and historically low-paid jobs. Even in times of high unemployment, the New York agricultural community can experience a labor shortage. Why? The common refrain among my interviewees was that those born in the United States were not interested in farm jobs, either because the work was too difficult, too temporary, or did not pay enough. A sentiment commonly shared across the nation was that no amount of pay was enough to recruit Americans; one Washington State apple grower, desperate for workers, told a reporter that he doubted that even $20 an hour would attract U.S. applicants.[45] Certainly the nature of the work sharpened the challenge of recruitment. Most farm labor is seasonal, and additional workers were almost always needed during the harvest, which could last as little as a month. Ask any farmer and he will tell you that it is next to impossible to get workers at peak harvest time if you have not recruited them in advance.

Until the late 1990s, locally based workers, including women and teenagers, were in the habit of filling some farm jobs, but in the early 2000s, I found that unless they had a specific personal link to a farm, they were unlikely to be hired or to seek out the work. Many of my interviewees reported that it was rare for native-born workers to apply for farm jobs. One apple grower said that he received only one or two calls a year from Americans, and when he described the job and invited them to the farm they did not show up. Echoing his peers, he told me that citizens might accept jobs as mechanics, carpenters, or tractor drivers but they never harvested in the fields. Farmers also repeatedly brought up the issue of reliability. Even if they were successful in recruiting locals or college students, they considered these workers to be unreliable. One technique that was described to me for assuring the reliability of one's work force was to pay on Fridays, like most employers, while at the same time using a Wednesday-to-Tuesday pay period. With this system the farmer "had three of their days," which would ensure that the workers would return on Monday.

Farmers also expressed clear preferences about the age of their workers. Almost every farmer I spoke to related stories of high school and college students who left at lunchtime and did not return or who lasted in the job only a few days. This situation is undoubtedly tied to the economy's move away from manufacturing jobs toward service sector work and the upgraded expectations that come with an increasingly college-educated citizenry. "Years back, everybody was workers," observed one grower. "Today, they are not." At the other end of the age spectrum, older workers—many of them black men, who seemed to be willing to stay in farm work through their seventies—were deemed by farmers to be too slow. Yet despite farmers' inclination to let go or to generally disfavor workers in their fifties and sixties, my interviews and research revealed a consensus among employers that black workers had aged out of farming and left on their own.

Hard work is one thing; farm work is quite another, according to a chronicle by journalist Tom Rivers about his own experiences in the farm fields of New York.[46] Rivers spent six months working one or a few days each month on more than a dozen farms. At the end of that period he ran his first marathon. Comparing the two experiences—both of which were new to him—he found the never-ending work in the fields to be much more arduous. There were none of the motivational cheers that inspire marathoners along the course, and he could see the toll imposed by the hidden nature of immigrant farmworkers' lives. For example, Rivers described meeting a twenty-two-year-old who lived "like a monk" and rarely left the farm, being too afraid even to go to McDonald's.

As for the physical nature of the work, Rivers recounts the pain in his hips, lower back, and shoulders that followed three hours of stooping while planting onions in freezing temperatures; some months later he picked peppers in the "scorching sun." He takes pains to describe the occupational hazards of using a razor-sharp six-inch knife to cut zucchinis and reports that he could barely grip the steering wheel at the end of the day because his hands were so tired. Ten hours of harvesting cabbages left him with shooting pains and meant that he couldn't pick up his twenty-three-pound, two-year-old daughter for the next four days. His wrists felt "mangled," but it could have been worse, as he was told the cabbages could be covered in snow later in the season. He likens three hours of harvesting cucumbers to being "trampled by wild bulls" and describes how it left him with throbbing wrists. Finally, after Rivers had spent six months working on various farms and was getting close to the marathon, he figured he was in good shape and hoped to keep pace with veteran apple pickers. At the day's end, however, he was stunned to discover that the best picker that day had picked 253 baskets of apples, earning $202, while the slowest had picked 198 for $158.60. Rivers had only picked 89, which would tally to $71.20 for eight hours.

A few growers acknowledged how grueling the work was, but most saw the imperfect labor market as the outcome of native-born Americans' reluctance to work hard, and therefore unrelated to the demanding work, poor remuneration, and precariousness of the work itself. Others were simply at a loss to explain the lack of a native-born work force. "It is curious that there is no domestic labor force," noted one of my interviewees. "There just doesn't seem to be any demand for agricultural jobs among American citizens, standard Americans, let's call them local American citizens." I did not ask for clarification on these terms, but given the context of our conversation, I took it to mean white Americans. In most low-paying sectors with a largely immigrant work force this is the case, and in agriculture especially there is a widespread perception that only Latino workers (and not South Asians or Africans, for example) are willing to work in the fields.

It was clear that most farmers did not want to hire the Americans who came looking for jobs and that they knew that their work force would consist largely of immigrants and guest workers who would accept the compensation offered. This is explained in part by dual labor market theory, which divides jobs in industrialized economies into primary and secondary sectors.[47] The primary sector includes stable, well-paid jobs that offer opportunity for advancement and well-established work rules, while jobs in the secondary sector offer none of those benefits and are governed by inconsistent work rules and poor working conditions. Dual labor market theory describes international migrants as uniquely motivated to accept secondary sector jobs since they are often target earners (i.e., looking to fund specific projects, such as home construction or their children's education) and do not seek out the social status that comes from primary sector employment.[48] Moreover, the conversion of wages from the standard of living in the United States to that of home countries allows the migrants to improve their families' quality of life in their home countries, sometimes in a substantial manner.

My interviewees had their own understanding of why no Americans would take farm jobs, yet at the same time they took pains to elaborate the benefits of a job. The same contradiction applied to the perks offered along with the job, such as free housing and access to farm-owned vehicles for workers' personal use. These perks would only be attractive to workers who were in need of shelter and who did not mind living on a farm in what might be considered substandard housing. One grower described his offerings in what he thought were desirable terms: "We pay for housing and all utilities. If you work forty hours, that's $250 or $275 a week, and all you have to buy is food." In addition, his father-in-law commented that employees also received produce from the farm. However, a local resident who already had a home would not be able to pay rent and maintain a vehicle on those wages alone.

Some farmers—particularly organic ones—relied on interns and apprentices. In a memoir about Hudson Valley farm life, businessman-turned-organic-farmer Keith Stewart explained that he relied only on an "apprentice workforce" of Americans who were paid a stipend "that is somewhat less than migrant workers receive."[49] He acknowledged that his peers hired migrant workers, usually Mexicans, but he had "thus far declined." One way to locate such apprentices was through World Wide Opportunities on Organic Farms (WWOOF), which linked certified organic farmers to job seekers. In the Hudson Valley, I was told, young Europeans and Israelis as well as native-born Americans could frequently be found "WWOOFing." The apprentices and WWOOFers were not simply farmers in training; they also included a significant segment of young people evincing a newfound interest in understanding where their food comes from. Reflective of a generation immersed in environmental ideals and values, they were youthful eco-entrepreneurs who wanted to get in at the ground level or just get their hands in the dirt for a few months. One WWOOFer I met, who had worked for a few weeks on New York's Long Island and for a month in California, described it as the cheap "hipster vacation" and an opportunity to see new places and meet new people. She only paid for her round-trip transportation, and while her labor contributed to the functioning of each farm, she did not imagine that she was replacing a hired laborer. Apprentices and other young workers, however, filled only a tiny portion of Hudson Valley farm jobs, and while these populations did offer an alternative work force, they were unlikely to provide the stability over many years that farmers have found with their immigrant workers. Stewart and many farmers I spoke to mentioned the constant effort needed to recruit and train young American employees who nevertheless did not stay long on the job.

Immigrant Workers

Immigrants from Latin America constituted the vast majority of the farm work force in the Hudson Valley. For the most part, farmers agreed that Latinos were a good match for the jobs on offer. Yet maintaining this work force was not without its challenges. Many interviewees complained about what they saw as punitive immigration policies, which made it difficult for them to hire a secure work force. They expressed concern that their workers might be arrested and deported, and that such an unfriendly climate would deter others from coming to New York. This was certainly the case in Western New York, where immigration authorities were particularly active[50] and the farmworker advocates I interviewed reported that arrests were made on a daily basis. I heard of farmworkers being stopped by immigration officials outside grocery stores and laundromats. Paul Baker, the former executive director of the New York State Horticultural Society, wrote an editorial in a grower newsletter arguing that border security was affecting the labor market and that labor shortages were an everyday occurrence.[51] Farmworker advocates also told me they were careful to have a lookout present during meetings with farmworkers, and that efforts were made to ensure that meeting places had more than one exit in case the undocumented needed to slip out. Although the Hudson Valley is a few hundred miles from the authority of the border patrol, employers were still concerned that its reach might be extended into their neighborhoods. Farmers were less preoccupied with immigration sweeps, but rather they feared that a routine stop by the police, such as for running a stop sign, might lead to the loss of workers.

I pushed most farmers on the issue of their workers' legal status when they told me that they hired Latinos. Their responses were clearly rehearsed, as reflected in the reply, "My paperwork is in order. My I-9s and my W-4s are done. The workers provide information. We cannot ask for social security cards by law; they just have to supply the number. They give us social security cards and green cards; the information is all there." Another respondent said he wasn't sure who was legal and who wasn't, claiming, "I know some of them have driver's licenses. I think some of them are legal." By law, employers must take workers' identifying documents at face value, but very often farmers could guess their workers' status based on other indicators, such as a complete lack of English-language skills, their mode of dress (newcomers wear give-away items like cowboy boots and hats), a lack of eye contact, and apparent inexperience with the culture of the hiring process. In addition, it was wide-spread knowledge that most Latino farmworkers were undocumented.

When I pursued the topic of whether farmers knew about the legal status of their workers, I experienced a range of reactions. One farmer declared that he found out his workers were undocumented only when they described how some fellow travelers died in the desert crossing the border. Uncomfortable with this knowledge, the employer successfully petitioned to get the workers to return to his farm as guest workers. He told me, "When I picked them up they were rested, they still had some of their winter fat, and were really pleased not to carry a one-gallon jug through the desert." In another case a farmer told me, "My H-2A [guest] workers are Mexican. A lot of them have been here for years, and whether their papers were good or not, I don't know. Maybe they let them lapse." This didn't make sense because employers have very specific contracts for individual guest workers, so I pressed him further. "I think everybody knows that some of them are illegal," he responded, but he continued, "I would be willing to say at least 75 percent of mine are legal." Based on my interviews with his workers, this was quite an overestimate. More often than not, the topic of whether or not one's workers were legally in the country drew either an animated response or complete silence. Most farmers did not find it an easy topic to discuss. In contrast, my employee interviewees were eager to discuss their legal status because they were looking for advice on how to change it.

Another pressing issue for farmers was the attitude of local populations toward their workers. With or without papers, the new immigrants were rarely welcomed, and community members tended to blame the employers for the "foreign" presence in their midst. Trask recounted that some locals could be "a little nasty" and used terms like "wetback." Others, he mentioned, realized that new immigrants were the only ones who would work on farms. "I have to be careful," he told me. "With the current state of the economy and people out of jobs, they are looking at my guys—some are driving decent cars and they are living for free and pulling a decent wage—and they are making comments to me." In hard times, the hiring of foreign workers raises questions about whether growers are making an effort to create jobs for those already living in the community. Workers isolated in labor camps had a lower public profile and less daily interaction with locals. But it might take only one angry local to bring down an immigration raid,[52] a dire and much-feared scenario even in locations that were hundreds of miles outside immigration officials' border jurisdiction.

American Dream

Debates about immigration cannot fail to be infused by time-honored narratives about foreigners who came to the United States and were first low-wage workers before they rose to prominent positions in business and society, thereby living out the "American dream." Given the stark reality of farmworkers' working and living conditions, as well as their exclusion from labor law protections, their legal status, and their limited future earning potential, the invocation of this archetype would seem inappropriate. Yet growers and industry advocates had their own, updated version of the "American dream," but rather than being lived out on U.S. soil it was considered an ideal to be pursued back in their workers' home countries. Consider a 2003 memo from the New York Farm Bureau that argued that the state's farmworkers' wages supported "themselves and their families in the country of origin." Despite the risks of illegally crossing the border and working with fake documents, wages in New York were better than they were in other countries, and this, according to the Farm Bureau, amounted to farmers "truly providing an opportunity for these individuals to live the 'American' dream."[53]

Workers rationalized their situations in the United States by comparing themselves to their counterparts at home rather than to other U.S. workers. Growers tend to operate under the same assumption, comparing the conditions under which their workers labor with the conditions in their home countries; comparisons to U.S. workers, especially citizen workers, are considered irrelevant. "Looking from an American point of view, it's a low wage," one grower observed, while also pointing out that a Guatemalan schoolteacher, for instance, might earn only $2,000 a year and a doctor maybe $10,000. A former apple orchard manager told me that his workers enjoyed a life of affluence in their home countries because of the hourly pay they received in New York, "They're rich," he told me. "I wish I had it as good as them." Another grower was more measured in her assessment of the benefits to her Jamaican guest workers, saying, "It's a good deal for them, except they are away from home."

It is part and parcel of American ideology that the exploitation of first-generation immigrants for the sake of future generations is justified. However, the persistent and mostly cyclical migration of Mexicans to the United States over the course of the twentieth century has resulted in a durable poverty. More to the point, the ability of immigrants to assimilate into the United States politically, economically, and socially is very limited today. Barriers to citizenship are much higher, and hostility from whites, although varying from decade to decade, is far from declining. Yet my interviewees found it easy to liken the situation of today's immigrants to those of yesteryear. One grower of Italian descent told me that years after his family bought their farmland, he heard that they had been "checked out" by neighbors for criminal or mob connections since in the 1960s Italian entrepreneurs were rare in his small Hudson Valley town. Evincing sympathy for his own workers, he told them, "I'm no different from you; I'm only three generations ahead of you. My grandfather went through this after going through Ellis Island." Others also typically observed that the situation of their workers was not very far removed from that of their own forebears. As one grower responded to an upstate reporter, "Maybe this is no different than our ancestors coming in here—just history repeating itself."[54]

Farmers also point to their workers' hometown projects, which have led to mobility at home, as proof that workers are well served by their employment on farms. In one instance a grower told me that workers on his farm had paid for electricity to be brought to their small Latin American town. After viewing a video that recorded the conditions in their town, he concluded, "They were living the way people here were living one hundred years ago, in shacks, no running water, no septic system." Another showed me photographs of a large church his workers had built in their hometown in Mexico, paid for with wages earned on his farm as well as the wages of others from the same hometown working elsewhere in New York. One of the New York State Department of Labor representatives I interviewed told me about a grower who visited the homes of his workers in Jamaica and found them to be among the most desirable in town. I was told this was a source of great pride for the grower and evidence that the system was of great and lasting benefit to the workers.

Although low-wage migrants may be economic heroes at home, this does not alter the fact that they are exploited in the United States.[55] Growers and Farm Bureau officials alike are all too aware that only the most poverty-stricken stand to benefit, in relative economic terms, from seasonal farm wages. Despite the well-fueled public perception that farm jobs are undesirable to native-born workers because of the hard work involved, it is the low compensation and difficult working conditions that are decisive deterrents to domestic workers. A 1951 federal report on this point is just as relevant today: "We depend on misfortune to build up our force of migratory workers and when the supply is low because there is not enough misfortune at home, we rely on misfortune abroad to replenish the supply."[56]

The iconography of immigrants as economic heroes at home is less grounded in reality when we consider that farmworkers in New York today are much less transient than previous generations of workers. Tighter immigration controls result in more immigrants settling for years at a time or opting to stay indefinitely along with their families.[57] Forty-one percent of the workers I interviewed lived in the Hudson Valley year-round. For those who remit as much of their wages as possible to family at home, life in New York is one of austerity. As workers settle in New York with their families and continue in agriculture, they join the working poor in the United States. If noncitizen Latinos continue in farm work and settle year-round in New York's rural towns, it remains to be seen if or how the rhetoric will shift to rationalize the use of these noncitizen domestic workers who have to survive within the U.S. economy.

Americanized?

Despite claims that work on New York farms offers an opportunity to achieve the American dream, farmers combatted such a message by clearly giving the impression that they did not want workers who were assimilated into U.S. culture. The downside of hiring Latinos, one farmer told me, is that if they stay too long in the states, "they get a little too Americanized. That's what I call it; they get Americanized, and then they get lazy." I heard about an employee on another farm who left for a while to do construction work and then returned. In his employer's mind, "that gave him an attitude" about his work since he knew he was skilled enough to find another job. Similarly, although immigrants' acquisition of English-language skills might ease communication with the boss, at least one farm owner opined, "You don't want Mexicans speaking English, because as soon as they start speaking English they start working like Americans." Another gave an uncomfortable reply when I asked who sold his goods at farmers' markets. "Americans, regular people I guess," he responded. When he recognized the implications of what he had just said, he was plainly embarrassed, so he repeated his answer but left out the "regular people" part.

Not every farmer I interviewed shared these views. One felt very strongly about not hiring the undocumented. His one Latino employee, Martin, was a green card holder. Unlike his peers, he would have preferred that his worker speak better English. Someone told him that Martin understood only about half of what was said in English, which shocked the employer. He had expected Martin to have greater competency in English, though he understood that he might not be fluent. On farms larger than his, which had only a handful of staff, farmers and workers usually relied on a bilingual worker to translate, and sometimes such a worker was promoted to the position of worker-supervisor due to his language skills. Ultimately, it seemed that communication between employers and Spanish-speaking employees left a lot to be desired on both sides. Some farmers made the effort to learn some Spanish, but most were dependent on the English picked up by their workers.[58]

Recognizing the Workers' Plight

Some farmers I interviewed expressed anguish about the trying circumstances of their employees, and such feelings served to cast them as sympathetic characters who were forced to hire the undocumented because no one else wanted the job. Trask, for example, commiserated with one of his workers who had not yet met his youngest child, telling me that such a situation was "not right" and "something needed to be done," placing blame on immigration policy. At the same time, farmers like him were businesspeople focused on minimizing costs, and so they did not always willingly make the connection between workers' needs or interests and the labor conditions they offered. Consider the fruit and vegetable grower who told me that farmers actively hired marginalized workers who had few alternatives for other kinds of employment: "It's not necessarily the workers' choice or something they are happy about," he explained, describing "abusive policies and horrible living conditions" that could result from their vulnerability. "It is something to be addressed with laws," he continued, "it is a policy issue. We've been up in arms and concerned about the overtime bill." The abrupt shift in his message made sense to him at the time, yet those last two sentences could not be more contradictory. He had set forth strong opinions about farmworkers' exploitation, and yet his call for remedies segued seamlessly into a harangue against a legislative proposal to award overtime pay to farmworkers.

Almost every interviewee referred to the camaraderie they shared with their workers, and all attested to their personal concern about their workers' livelihoods. Yet running a business required that they think about their employees in terms of the expense and quality of labor inputs. And the prejudices of race and class so deeply inscribed in the American culture they inhabited also shaped their view of these poor immigrants. Only one of them supported overtime pay for agricultural workers. He recognized that the extra cost would put some farms out of business, because "whatever you do to give more to one takes away from the other." Even so, he firmly believed farmhands deserved as much as other workers. His was one of the few interviews that acknowledged farmworker conditions as a matter of justice. Take justice out of the picture, he argued, and we would continue to undervalue the labor of farmworkers and other immigrant workers by paying them as little as possible.

My interviewees' opinions everywhere revealed the contradictory opinions they held about their workers. On the one hand, they needed to squeeze whatever they could from their work force at the lowest possible wage. On the other hand, many evinced sympathy for the many problems faced by their employees. These contradictions played out in many ways. Generosity on the part of the employers might also appear as the height of paternalism. Employers who saw themselves as offering an escape from third world poverty or as protective bulwarks against community resentment could not avoid taking advantage of workers' undocumented legal status to tighten their control over them. When farmers likened the precariousness of their own situations and their hardscrabble family histories to that of their workers, they displayed some recognition of the workers' plights, but at the same time they downplayed the differences in risk and opportunity. It was not always easy to disentangle authentic sympathy from self-interested sentiment, and indeed they often seemed inseparable.

FIGURE 1. Harvesters of summer vegetables survey a field.

FIGURE 2. Workers pull weeds from an onion field.

FIGURE 3. Workers pick and tie greens for local distribution.

FIGURE 4. The squash harvest requires repeated bending and using a sharp knife to cut the stalk.

FIGURE 5. Harvesting leeks involves trimming their long leaf sheaths.

FIGURE 6. Band-Aids don't stand up to the work routine, so here a worker has used duct tape on his cuts.

FIGURE 7. Freshly harvested sweet potatoes are packed to go to market.

FIGURE 8. Packinghouse workers inspect and pack zucchini.

FIGURE 9. Four undocumented mothers display electronic tracking ankle bracelets used by U.S. Immigration and Customs Enforcement (ICE).

FIGURE 10. Workers and advocates call for equal rights for farmworkers at the annual Justice for Farmworkers Day in Albany.

FOUR

Sustainable Jobs?

ETHNIC SUCCESSION AND THE NEW LATINOS

IN THE LAST DECADE OF the twentieth century, the composition of the agricultural work force in New York State changed dramatically. The number of black farmworkers, including African American and Caribbean, declined from almost half to one-quarter of the work force, while the presence of Latino workers increased from one-third to two-thirds.[1] This was a shift not only in race, but also in legal status, from citizen and green card holders to the undocumented. In the ensuing years, the black farm work force, including migrants from the South and local settled-out workers (those who move to New York after being migrant workers), has continued to shrink. In sharp contrast to other agricultural regions across the United States, New York's farmworkers were primarily African American for the better part of the twentieth century, and they were joined by West Indian workers, whose numbers started increasing in the 1970s.[2] These West Indians included guest workers, as well as former guest workers who overstayed their visas. The overall population of black workers began to decline, however, during the 1960s, as mechanization displaced many migrant workers and the number of New York's seasonal workers dropped by half.[3] Other factors that decreased their number included urban migration and increased job opportunities in southern states.[4] Moreover, few farmworker children followed their parents into the fields, as they received a better education than did previous generations. Certainly another factor was the rapidly increasing migration of Latin Americans all over the United States starting in the in the 1980s.

On the face of it, the demographic shift from African American to West Indian to Latino laborers appeared to be a neat transition from a population that was no longer interested in agricultural labor to one that had fewer economic alternatives. None of this happened naturally, however. As I will argue in this chapter, agricultural employers, ably assisted by state agencies, actively managed this ethnic succession by ushering into the region new immigrants in order to secure a tractable work force. The new immigrants, the vast majority of them without papers, were preferred because of their work ethic and because of their reluctance to complain about their work circumstances, both of which stemmed from their vulnerabilities. How else do we explain the rapidity of the ethnic succession on New York farms, especially in the shift from black to Latino workers in the Hudson Valley itself? In the pages that follow, I offer examples of farm-level hiring decisions that will illustrate how the transition was achieved and contextualize them within the regional pattern of the industry's labor market. In particular, I will focus on the specific role played by the state Department of Labor in substituting Latinos for black workers. Finally, I will show how employers tried to justify the succession by disparaging former workers and extolling the virtues of the newer ones. These characterizations tried to mask, but were rooted in, the racialization of different groups of workers, and they often obscured the agency of workers who were beginning to utilize their labor power to protest and improve their work conditions.[5]

This discussion is particularly relevant to the local food movement, which has been widely praised as an engine for economic revitalization, in part due to the jobs it is helping to create. The new jobs are not only in a field defined by environmental sustainability, but they are also held up as sustainable employment. These jobs, it is assumed, are more likely to endure because the local networks of consumption on which they are built will not be affected by competition from other regions, whether in the United States or overseas. Moreover, the promise of sustainability will surely draw in a new generation of U.S.-born recruits, hungry for work that is relatively secure and morally gratifying. In the course of my research, however, I met only one U.S.-born worker engaged in the fields or packinghouses. In addition, employers' opinions and hiring practices clearly demonstrated that they did not want to hire local or American workers. Although farmers today insist that they cannot find American workers, this chapter shows that, until quite recently, there were American workers (including green card–holding immigrants) employed in significant numbers on Hudson Valley farms, and that farmers actively recruited undocumented Latinos to replace local workers and American migrants who were becoming more demanding.

This case departs from prominent theories explaining increases in immigration, such as neoclassical economics,[6] dual labor market theory,[7] network migration theory,[8] and scholarship on new immigrant destinations. These theories generally do not account for the agency of employers in actively promoting immigration. Supply-side arguments also tend to rely on macro explanations of immigration, such as the impact of free trade agreements.[9] Although an understanding of migration networks is vital for comprehending the process of immigration, the scholarship tends to focus on the supply side, often with a methodological scope limited to an analysis of actors in the sending country, and tends to exclude a variety of players in the receiving country who actively participate in the promotion of migration streams.[10] One group that clearly requires study is domestic employers, but they are largely absent from immigration literature.[11]

Recently, scholars have focused their attention on "new destinations," such as Omaha, Nebraska; Marshalltown, Iowa; Dalton, Georgia; and other locations that have seen an exponential increase in Latino newcomers.[12] With regard to the geographic and historical specificities of increased immigration to new localities, it has been posited that the confluence of several factors in the 1990s drove immigrants away from traditional receiving areas. California and Texas, along with the rest of the Southwest, once received the largest number of Latino immigrants, due not only to their proximity to Mexico, but also to employers' historical reliance on Mexican workers. Yet the downward turn in the California economy and job market, growing nativist sentiment and anti-immigrant policies in California, the increased militarization of the border, which made it more difficult to illegally enter California and Texas from Mexico, and the availability of legal documents as a result of the 1986 amnesty program made better jobs and alternative destinations more available to the newly legalized.[13] The 1986 Immigration Reform and Control Act (IRCA) legalized 2.7 million undocumented persons under its amnesty provisions.[14] However, none of these factors would help explain the influx of Mexicans into agricultural work in New York State. In this case, then, although New York represents a "new destination," it is unique for not conforming to scholarly accounts of increased Latino immigration at the turn of the twenty-first century.

FARMERS MOVE FROM BLACKS TO LATINOS

Bobby Trask's farm business is a thriving concern today, and in the course of our interviews he shared his path to success with me. Twenty years ago he was weeding his own vegetables and struggling to figure out a business plan. Like many farmers I met, he credits part of his success to his workers, saying, "I couldn't be doing what I'm doing right now if it weren't for my crew." But this was not always the case. "It was getting crazy," he recalled. "In most of the '80s it was sporadic. You had a work force, [and then with] fights and drunks, people would leave. There was consistency and then there wasn't. You would lose guys, call the labor department for more, and then you get people showing up when you didn't need them. Then you would be out of guys again." When the farm was very small, he and his father employed local youth, but by the 1980s teenagers were showing much less interest in farm work, so Trask bought a trailer and began to recruit black Americans from Florida. They were "good workers," he recalled, but "blew all their money over the weekend on booze and women and asked for advances come Monday." He explained how he then switched to hiring Puerto Ricans, whom Trask described as "better workers," but the drinking continued along with the occasional knife fight, plus the newcomers did not get along with the black workers. After Trask had to call the state troopers once too often, he switched workers again. This time it was Mexicans whom Trask found "more productive and generally more focused," but when they quit during his busiest season, he was left in a lurch and uncertain about the farm's future. Local workers were not an option, he stressed, because they simply did not want the work. He had tried hiring college students who were home for the summer, but they could not staff the rapidly expanding farm and were only available in the middle of the farming season. High school students, he recalled, showed some initial enthusiasm for farm work, but that quickly wore off, and in any case they required too much hands-on managing.

Someone recommended that he go south to Westchester County to recruit landscaping workers. After waiting in frustration for an hour and a half, he was rewarded with one Latino worker. Trask brought him back to the farm, ranting to him for the whole drive, Trask says, about the scarcity of workers that hurt his business, even though his passenger understood no English. Yet that single employee would help transform the entire business as he went on to recruit family and friends, first from Westchester County, then directly from his rural hometown itself. Over the next two decades dozens of Latinos would arrive to staff the farm. "Some work for two years and leave for two years," Trask reported. "It seems like I have this revolving thing going on. But I've got some guys who have been with me going on eighteen to twenty years." The change in the durability of his work force was dramatic. "My turnover went from 70 or 80 percent to next to nothing. I was able to just go out and work."

Marty Crane, the farm manager at an apple orchard and packinghouse operation for the past twenty years, also discussed his hiring strategies with me. He hired only Latinos, about seventy-five of them. Twenty-five were employed from August through March, fifteen worked almost year-round, and the remainder arrived only for the apple harvest. He hired mostly Mexicans and Guatemalans who lived on the farm or nearby; most of them staffed the packinghouse, though some were also field workers. At the busiest time—during and right after the harvest—the farm employed Guatemalan guest workers in the packinghouse. Additional apple pickers included Mexican guest workers and undocumented Mexicans who traveled the migrant stream, harvesting Florida oranges and New Jersey blueberries before arriving in New York, whose wives picked up work in packinghouses. Crane also applied through the state government for Jamaican guest workers as a backup. In the case of a labor shortage, he explained, the Jamaican government would be very efficient in providing workers, but he had not employed them for years as he has not needed the extra workers. One reason for the diverse composition of his work force was that Crane wanted to be sure that some of his workers were in the United States legally: in the late 1990s, he had lost quite a few workers when immigration authorities came to the farm.

Hiring different groups of workers or different crews, while more challenging for management, assured some level of security. Trask had lost a crew who walked off the job, Crane had lost workers to deportation, and I heard from other employers and workers about cases in which an entire crew of migrant harvest workers would disappear overnight. Employers described being flummoxed when this occurred. Workers who told me of two or three such scenarios reported they occurred as a result of a disagreement with the boss over wages or conditions, and remaining workers often had conflicting stories of whether workers were fired or left on their own. By hiring different groups, the employer could try to prevent the development of worker solidarity, particularly if workers were segregated by nationality and legal status, in the job tasks they were given, and in housing, which was common. In this way, the departure of a group of disenchanted workers would not unduly affect the work force as a whole.

Cy Valen, an older farmer I interviewed, recalled when blacks first arrived for farm work during World War II. His family continued to hire African Americans, including a few families and a crew of single men, until the 1980s and then switched to Jamaicans. In 2007, when his son-in-law Ronnie Patten took over the farm's management, he began to hire Latinos even though Valen still had a preference for Jamaicans. Patten conceded that the blacks handled the machinery better and fewer repairs were required, but the former, he said, were too old and slow, and he "got tired of watching them sit around." Patten pointed out that the black workers had aged out and the young ones did not want to work. Valen disagreed. To find workers Patten initially called the state Department of Labor and other farmers to find Latino workers, but by 2009, when we met, there was no need to tap into those connections. "I get extra labor directly from my workers. They know this one or that one," he said.

When I interviewed the Jamaican guest workers at an Ulster County apple farm in 2002, they told me that every year they worried it would be their last, as fewer were invited back annually. The grower had informed them that the Mexican workers were much cheaper and that it was no longer cost-effective for him to employ them. Those I spoke to were in their late fifties and early sixties. From a certain perspective, it might be said that these workers were preparing for retirement. Indeed, the termination of older workers is often rationalized by claiming that they "aged out" of farm work. However, it was clear from our conversation that they were willing and hoping to return to New York annually through their early seventies. In addition to "retiring" some of the guest workers early, their employer, I was told, was giving the Mexican replacements more hours, including some rainy days when the Jamaicans were told that no work was available. When I visited again six years later, the farmer was no longer employing any Jamaican guest workers.

Another case from Western New York, outside the Hudson Valley, illustrates the transition in a different way. The Marcuses owned an apple orchard that had been in the family for several generations and had been staffed by southern black migrants for as long as they could remember. In the 1980s, they hired Trinidadians based in Florida in addition to U.S.-born blacks; a decade later they shifted to U.S.-based Jamaican and Haitian workers. Although I could not determine the legal status of these workers, I found in my research that older Jamaican workers often had a green card, invariably obtained after they left the guest worker program and married U.S. citizens. Younger Caribbean workers, including Haitians, tended to be undocumented. Starting in the early 2000s, the Marcuses began to employ one year-round Mexican worker, and around 2004 they made the decision to shift to an all-Mexican work force. They had moved away from growing apples for processing toward fresh fruit production. That change, they told me, required "higher standards," and labor quality was more important. Their perception, shared by most of the farmers I interviewed, was that Latinos were more diligent and dedicated workers.

The decision to go with an all-Latino work force was also spurred by problems they had experienced with their Haitian crew who migrated from Florida for the harvest. "Times and attitudes change," one of them remarked. Although the Haitians were good pickers, the Marcuses described them as "rebellious by nature" and harder to work with (this was a characterization of Haitian workers that I heard repeatedly). The Haitians they employed as tree trimmers, for example, tried to negotiate a higher wage almost every day, whereas they found the Mexicans to be less demanding. Moreover, some of the Haitian workers expected in New York would not show up but would send new workers in their place. Of the fifteen Haitians who arrived in 2004, only five workers had previously been employed by the Marcuses, whereas in the past, at least twelve of the core group would return each year. They told me that the Haitian workers were finding full-time employment elsewhere and that the ones who did come to the farm did not speak English and wanted more hours than were available. Consequently, these workers always seemed to be "angry" and complained persistently about not having more work (2004 was a particularly rainy year, which meant a lot of days off). According to the Marcuses, the Haitians also became careless at picking, which jeopardized the farm's profits. Seen from the workers' perspective, these patterns of behavior were likely to have been an effort to show their discontent or express their grievances and thereby to achieve better wages.

Because the Marcuses did not speak Spanish and because of the structure of farm work in Western New York, the new Mexican workers would be managed by a crew leader rather than directly by the Marcuses. In general, contractors and crew leaders are more common on Western New York farms than in the Hudson Valley. On one hand, it may be a relief for the employer not to be in the fields constantly, but, on the other, it would mean a loss of control and a loss of profits since supervisors are paid more than workers. Also, while the Haitians had been legal, and the Marcuses felt that most of the Mexicans had "good ID" (which I took to mean false papers or green cards), they were worried about losing their new workers due to their legal status. Switching to guest workers, however, involved a lot of paperwork, and after 9/11 the application procedure had become even more arduous. The

Marcuses were not alone among the farmers I spoke to in expressing impatience with the process of procuring foreign guest workers.

Other growers helped the Marcuses make their decision to switch to Mexicans by sharing their positive experiences and recommending a particular labor contractor who would provide them with a crew and crew leader. The Marcuses expressed remorse about breaking off ties with their previous workers, who would be informed by mail that the jobs were no longer available. But the core group that returned every year was getting smaller, and the Marcuses no longer knew the crew all that well. Ethnic succession on farms like theirs was closely managed by the growers themselves. The Marcuses were switching from a work force they could no longer easily control and who complained they were not earning enough to one that made few demands at all. In this they were following the example of their peer growers who had experienced more "success" with a substitute Latino work force.

James Sweet, a Dutchess County fruit and vegetable farmer, claimed he was reluctant to hire undocumented Latinos and instead continued to rely on Jamaican guest workers. Undocumented workers could be taken away, and he had also heard stories "where the crew leader doesn't get along with the grower and just ups and leaves." Guest workers, he pointed out, "are not free to leave," acknowledging that "this might not sound so good, but it's really important because our work is so time-sensitive." Sweet had hired one Latino worker, whom he heard was looking for farm work, and helped him secure a green card—a process he found adversarial, intimidating, and frightening for his worker, who was seeking asylum. His son told me, "It's not cost-effective to always hire legal laborers, and there's a lot of hoops to jump through with the H-2 [guest worker] program. But we've never employed an undocumented laborer here, and Dad is really proud of that."

These examples illustrate the rationale of specific growers, but the overall outcomes reflect a systematic trend away from the use of migrant and settled-out blacks; African Americans are all but gone on Hudson Valley farms, and the number of West Indian workers continues to decline. Moreover, the guest workers arriving in New York are increasingly Latino instead of Jamaican.[15] Notwithstanding the range of rationales offered, farmers' gravitation toward hiring Latinos represents a longer pattern of hiring the most vulnerable and compliant workers to staff their operations.

Farmers offered job access to preferred workers and limited it to those less favored. In this way, they were effectively able to manage the ethnic succession of their work force. As I have shown, one common hiring technique of the growers was to get staffing recommendations from currently favored workers. Once they had hired some Latinos, growers could effectively control further recruitment by relying on the kin and community ties of these "pioneer workers." The acquisition of jobs through such networks was evidenced by the fact that 76 percent of the farmworkers I interviewed in 2002 who were born in Latin America came directly to New York from their home country rather than working in another state first. Such a statistic distinguishes this case from other instances of "new destinations," where secondary migration from other U.S. locations, particularly California, was so significant.

Family and friends were an extremely important factor in enabling workers to find employment in the Hudson Valley. Indeed, four states in Mexico, Hidalgo, Puebla, Oaxaca, and Querétaro, accounted for almost half of the workers I interviewed. This kin system is quite distinct from farm labor contracting, in which a labor agent acts as the primary link between immigrants and jobs. I identified contractors on only a few of the farms where I conducted my interviews, although they were much more numerous in Western New York. Sixty-nine percent of those I interviewed reported that they heard about their job through family and/or friends, and this figure rose to 85 percent for undocumented workers. Furthermore, 75 percent of all workers and 89 percent of non–guest workers (i.e., undocumented, resident, and citizen workers) worked in the Hudson Valley with family and/or community members from their home countries. Relying on kin ties was a straightforward way of managing recruitment, but what about the grower who had not yet made the shift to a Latino work force? After all, depending on kin ties requires having at least one worker who can link to others like him. To hire their first Latino worker, some growers traveled. Three workers from the northern Mexican city of Juárez had actually met their New York boss in Mexico, where he had gone to recruit workers, and I heard of another grower who recruited his Mexican workers during a trip to North Carolina, where a family member had a farm.

On some farms the ethnic shift had been slow and piecemeal, while on others it happened very quickly. Crane detailed the changes he oversaw in the farm's work force from the time he was hired in 1991. At that time all the field workers were Jamaican guest workers. In the packinghouse there were settled-out and migrant black workers (mostly Jamaican) and a few older white women, but about half of the packinghouse workers were Mexican (the first Mexican arrived on the farm around 1988). Nine years later, the farm had no black workers at all. At first Crane continued the farm's practice of employing Jamaican guest workers each year, some of whom had been returning annually for fifteen to twenty years. But because he had successfully worked with Mexican migrant workers before coming to New York, he made a conscious choice to shift the work force. Initially he put in a call to a crew leader he knew from another state and asked for workers. After that, those he hired introduced him to their kin, and each year he hired more. He also began to train Latinos to work in the orchard, whereas previously they had worked only in the packinghouse.

THE ETHNIC SHIFT IN HISTORICAL CONTEXT

From World War II through the early 1970s, the vast majority of migrant workers in New York were African Americans from the South. This was a labor market profile relatively unique to the state and was uniformly evident, whether on Long Island potato fields, Hudson Valley fruit and vegetable farms, Wayne County's apple orchards (on Lake Ontario), Western New York's bean fields, North Country dairies, or the Finger Lakes vineyards. In 1960, New York was employing 27,600 interstate farmworkers who were almost exclusively African American migrants (including youth).[16] These blacks were considered New York's "traditional" migrant workers.[17]

In the Hudson Valley in the 1970s, African American farmworkers were largely supplemented and then succeeded by West Indian laborers, mostly Jamaicans. Haitians fleeing Duvalier's oppressive policies also had a presence on New York farms in the 1980s, before Latinos came to dominate the regional agricultural labor market.[18] In this sense, the dominant pattern of ethnic succession in the region is from African Americans to West Indians to Latinos. Consequently, the employee base went from citizen African Americans to Caribbean guest workers. These guest workers became undocumented and then earned green cards by staying in the United States and marrying U.S. citizens. Later the shift was to undocumented Latino workers. Another way of describing this trajectory is the shift from native-born citizen workers to a noncitizen immigrant group that largely earned legal status, and then to another noncitizen immigrant group that has remained undocumented.[19]

The profile of guest workers has also changed in the last two decades, from Jamaicans to Mexicans. For the past several decades, New York growers have hired Jamaican guest workers primarily to work on apple orchards during the harvest. The number of guest workers was around 2,000 annually until 2006, when it started to rise, reaching 3,000 in 2008—a small portion of the overall farm work force, which is estimated at between 30,000 and 60,000.[20] This recent increase in guest workers may be due to labor insecurity, because undocumented workers are at risk of deportation, particularly in Western New York, and farmers want to be sure they have laborers at the critical harvest time.[21]

World War II generated a labor shortage on New York's farms that stimulated the hiring of southern blacks and also led to the British West Indies Program, a guest worker program that corresponded to the West Coast Bracero Program, both of which were considered emergency labor programs. Most of the West Indians initially arrived on East Coast farms as foreign guest workers from Barbados, Dominica, Jamaica, and St. Lucia. In the 1960s, when the Bracero Program was shut down in response to human rights concerns,[22] the East Coast program continued as the smaller H-2A program, which supplied mostly Jamaican workers to New York fruit farms.[23] Each year, some workers would drop out of the program and simply stay on in the United States. Since most guest worker visas would expire within a year of arrival they became undocumented, but many would earn permanent resident status either through marriage or through the amnesty provisions of the 1986 Immigration Reform and Control Act. Already initiated into U.S. agriculture, West Indian workers who had left the guest worker program, like the African Americans before them, traveled the East Coast migrant stream annually and also settled out permanently in the United States, including in the Hudson Valley, where they were commonly employed on apple orchards.[24]

Puerto Ricans employed in New York agriculture deserve some mention. The U.S. government brought in wartime workers from the island to the East Coast in 1944; in 1948 a federal Department of Labor program called the Migration Division established government-sponsored recruitment and transportation of Puerto Ricans for New York jobs, including on farms.[25] Many of these workers established roots around the state and then facilitated the direct hiring of other Puerto Ricans. Compared to African Americans, they were small in number, and, despite their significant numbers in the farm labor markets of surrounding states, there are now relatively few Puerto Ricans employed in New York agriculture.[26] Among those I spoke with who provided services to farmworkers there is a widespread belief that Puerto Ricans are intentionally passed over by New York farmers because they are citizens and are therefore more inclined to challenge poor pay and labor conditions. Indeed, their history on Western New York farms provides evidence of their self-empowerment during the 1960s. Anthropologist Ismael García-Colón detailed a 1966 Western New York protest by Puerto Rican farmworkers that influenced a region-wide rise in complaints by other farmhands about their work, housing conditions, and community discrimination. This small-scale revolt lasted for several years. At that time, the Puerto Rican Department of Labor responded with a campaign to inform farmworkers traveling to the mainland about their rights and responsibilities. As a result, these workers pushed for the improvement of farm labor conditions[27] and helped establish the statewide perception that Puerto Ricans were difficult to control in the agricultural workplace.

Testimony from my interviews offered a nuanced description of workers' nationalities and also revealed that in the Hudson Valley the timing of the ethnic transition varied somewhat according to the type of farm and location. A service provider who used to be a farmworker in Ulster County recalled that in the 1980s farmworkers were about half African American and half Jamaican. By the late 1990s, however, the work force was mostly Jamaicans with a small percentage of African Americans, who were largely from Florida, Georgia, and Alabama. In 2010, however, Latinos accounted for 80 percent of the work force, he estimated. He argued that apple orchards were the last to transition away from black workers. In addition, one Jamaican laborer I interviewed recalled that in 1975, the year of his arrival in Ulster County, there were "no Mexicans, no Puerto Ricans, just pure Jamaicans." A migrant health service professional largely confirmed these accounts and offered a fuller description of the diversity of the regional farm labor market. When he started working with the farmworker population in 1989, the apple workers were mostly Jamaican guest workers with a "remnant" of African Americans. By contrast, the corn crop was staffed by African Americans from the South and a small number of Haitians from Florida. By 2009, though, all of these crops were primarily staffed by Mexicans. By his reckoning, those working on the onion crop had changed the least. In the early 1990s onion growers were mostly employing Mexican Americans and Mexicans, with an assortment of Central Americans, Puerto Ricans, and Filipinos. Yet, by the late 1990s, this mixed work force had become increasingly Mexican.

Beginning in the late 1980s, Latin American immigration to all parts of the United States dramatically increased, including to what scholars have identified as "new destinations," where substantial populations of Latinos did not previously exist.[28] New York City was a major gateway destination. Until the late twentieth century, the city's Latinos were primarily Puerto Rican, Dominican, Cuban, Colombian, Ecuadorean, and Peruvian.[29] The Mexicans that did live there were largely from the Mixteca Baja region, which includes parts of the Mexican states of Puebla, Oaxaca, and Guerrero.[30] This was equally the case in the lower and mid–Hudson Valley, particularly in the cities of Poughkeepsie (in Dutchess County), Newburgh, and Middletown (both in Orange County).[31] This Latino immigration to Hudson Valley cities was spurred by the promise of employment in urban service industries, but it was also encouraged by the availability of work on local farms. This helps explain why Latino farmworkers were more concentrated at an earlier date in the lower Hudson Valley than they were in Western New York.

The surge of Latino, mostly Mexican, immigrants to new places came on the heels of the amnesty programs of the 1986 Immigration Reform and Control Act (IRCA). Newly documented Latinos from the West and Midwest began to migrate east and establish what has been termed the "new destinations of an old migration."[32] In addition, IRCA included the Seasonal Agricultural Worker (SAW) provisions, which offered a path to citizenship for anyone who could prove they had worked at least ninety days as a farmworker in the three years prior to the act's passage. As David Griffith points out in his examination of changes in the agricultural labor market, some of those who earned legal status through the SAW provision went on to play a role in recruiting and transporting undocumented immigrants from Latin America to U.S. farms. This seemed to be a more common scenario in Western New York, where contractors were much more common. (Only 12 percent of the Hudson Valley workers I interviewed were employed by contractors.) Moreover, the amnesty program encouraged further undocumented immigration by sending the message that legalization was dependent on demonstrating work experience in the United States.[33]

Complementing these accounts, an oral history of Wayne County apple pickers conducted in the mid-1980s confirmed that the three thousand seasonal workers surveyed were mostly African American, Puerto Rican, or Haitian, with "an increasing number" of Mexicans.[34] Griffith estimated that the replacement of African Americans by Mexicans in Wayne County occurred "almost completely in roughly ten years."[35] As many of my own interviewees explained, the SAW provision also embraced West Indian workers who had left the guest worker program but had not secured legal status through marriage. While legal status was a ticket out of farm labor, some workers remained on the job because they were accustomed to farm work and lived year-round on farms. One former farmworker who qualified for amnesty through the SAW provision told me he stayed on for "a good period until I say I want to go better myself," at which point he secured a higher-paying year-round job maintaining the grounds of a golf course.

THE STATE FACILITATES THE SHIFT

Another common method of recruitment was through the New York State Department of Labor Rural Employment Program (NYSDOL), which offers employment services to farmers and prospective workers. Once the NYSDOL was informed of a job opening, its staff had to create a job order and post it online in a nationally accessible Department of Labor system called the New York State Job Bank. Several of the service providers and advocates I interviewed attested that the job bank opportunities were processed in a way that prevented the hiring of domestic, mostly black, workers. One provider reported that the website was not user-friendly and that it was very difficult for out-of-state workers to secure jobs through the system. She had seen agricultural jobs listed that required a resume, which in her opinion only served to restrict access to jobs. Her colleague reported that it was common for native-born workers to apply for such jobs online only to be told that they needed prior experience in harvesting apples. In the meantime, the same grower would be applying for foreign guest workers. As this service provider put it, "Where are the apple orchards in Jamaica and Mexico for those workers to gain experience?" His suspicion, widely shared among service providers, was that the job bank was used by farmers only as a convenient way of fulfilling federal requirements. Jobs had to be advertised and go unfilled for the growers to qualify for receiving guest workers.

Aside from the job bank, NYSDOL rural representatives were available to place workers who called them or showed up in person. These rural reps were very familiar with farms and workers in their region. And the more Latinos they know, the more the reps were introduced to other Latinos in these workers' networks, enabling the hiring of more Latinos. The job bank, according to service providers I met with, is used by a fair share of out-of-state American workers looking for work, but the rural reps were inclined to place workers who were already in New York instead of hiring out-of-state workers, thus facilitating the shift to Latinos. As one service provider put it, referring specifically to the rural reps' role in the ethnic succession of the work force, "U.S. workers are not sought out." The service providers received the same job requests through the job bank and would call the local rural reps with leads to workers in search of employment, but invariably there would be no response. "You can tell from the surnames which ones are Latino, and they will get work quicker and faster than African Americans," she explained. "It's because of the relationship that the employer has with the DOL. The growers tell them who they want. They say, 'Don't bring me those lazy people who are on welfare.' That message is very strong."

The NYSDOL's placement of Latino workers reflects not only growers' preferences, but also agriculture's peculiar hiring needs. When a grower realizes that he or she needs extra workers on very short notice, hiring is an urgent matter. When there is a need to place workers in jobs immediately, the rural reps have an incentive to tap into local networks (they often hear of new arrivals in search of work) instead of relying on the job bank website or other organizations that place workers, which often take longer and involve a more laborious process. Such network hiring also favors hiring Latinos at the expense of black workers. Yet growers insisted that they could not find locally based workers to take agricultural jobs, and virtually all my interviewees attested that American workers were not willing to do the hard work involved.

The employment of guest workers is a well-documented mode of labor control in the United States and other advanced economies, yet guest workers account for less than 10 percent of hired labor on New York's farms. When harvest time comes, growers always need extra workers, and the need can fluctuate daily according to the weather patterns. The vegetable harvest can last from spring well into autumn, but the apple harvest is concentrated into a much shorter period of time. It can be difficult to find workers for a two-month stint only, and this is why guest workers have typically been concentrated in the apple industry. Many of my interviewees had relied on guest workers for decades, and were in the habit of requesting workers by name in order to ensure stability from year to year. Increasingly, however, they were transitioning away from using guest workers and shifting to undocumented workers.

The U.S. government has an interest in not taking jobs away from U.S.-based laborers. On the face of it, the guest worker program is structured to prevent this from occurring more than it already does. Under this program, guest workers are paid an adjusted wage that is several dollars higher than the minimum wage. In addition, growers have to show that no U.S. workers are available. Growers are required to advertise these jobs in venues like the New York State Job Bank, but they must hire U.S. workers who apply for the job and pay them the higher, adjusted wage, even after the guest workers have arrived. One service provider explained that in many cases growers simply refuse to hire the predominantly black workers who show up. Yet they always seem to hire the Latinos, who are considered domestic workers when they have counterfeit documents (employers are required by law not to question birth certificates, social security cards, and green cards if they appear valid). As one grower told me, "I had times years ago when somebody told me one card was good and it wasn't, so I can't tell. They have holograms and stuff now, but they all look the same to me." In other words, the undocumented pass muster as domestic workers in a technical sense, even though almost all employers know that the vast majority of Latino farmworkers are undocumented. In a strange twist, at least one grower explained to his Jamaican guest workers that Mexicans were domestic workers, and so he could not continue to hire the Jamaican farmhands. As a former Jamaican guest worker recalled, "One farmer told me that he had to hire the Mexicans because of the government program—the legal thing."

In low-wage work there is a hierarchy of wages that corresponds to the workers' legal status, with the undocumented earning the least and those with green cards or citizenship status receiving better pay and working conditions.[36] When I asked my interviewees why they thought their employers did not hire citizens, they often referred to the perception that citizens are going to walk off the job, ask for a higher salary, or leave for the day if they don't feel well, and that they prefer to drive tractors or other machinery over performing more manual labor. The undocumented work harder for less money. As one interviewee put it, "The illegal workers, we don't have much of a choice. We need money. What other option do we have? It is the game of capitalism, you understand? In this society there is no equality for us. We are the slaves and the ones ruling are the kings." Similarly, when asked how the job might be different if he were a green card holder or a citizen, a Jamaican guest worker gave a typical response: "I could ask for more money. Maybe I could walk off the job anytime I was ready. I could get an easier job."

This scenario is likely to be repeated in the event of further immigration reform. The beneficial impact of previous amnesty programs on the undocumented has been substantial for those who earned green cards, enabling them to climb the legal status hierarchy and find better jobs with more rights. At the same time, however, the amnesty programs perpetuated the same categories for subsequent immigrants without this status. The IRCA, which offered amnesty to undocumented immigrants, sparked a wave of arrivals of unauthorized immigrants,[37] who were destined for low-wage, substandard employment in the fields, construction sites, restaurant kitchens, and factory sweatshops across the nation.

Several sources informed me that farmers and NYSDOL agents consider it a problem when native-born workers (including Puerto Ricans) apply for jobs. One remedy, which I was told came at the explicit, but unofficial, recommendation of the NYSDOL, was to close a job order if the "wrong" kind of workers applied and then reopen it later in hopes that the prior applicants had moved on. The reluctance to hire domestic workers was described to me by a grower who had a small crew of Mexican guest workers. In line with federal requirements, he had posted the jobs as available to workers in the United States. When someone in Arizona applied for the job, the farmer explained, "Almost all of my rights were gone. I pretty much had to hire him no matter what, as long as he met the criteria that was on the [job order] form." The farmer went on to complain that he could not conduct a background check on the applicant and expressed concern about having him live on the farm. "That was what was really scary to us. You don't know what this guy did in Arizona." This grower, it appeared, was concerned about the safety of his workers, who would have to make room in their trailer for a stranger, and he was pleased when the hire did not work out. Housing farmworkers is undeniably a complicated business, and the grower's concern for his workers reflected the sensitivity of this matter. Yet this tendency to *precriminalize* domestic workers also served as a convenient way to justify not hiring them.[38]

At a NYSDOL preharvest conference that I attended in 2003, a service provider who was moderating a panel engaged in a heated exchange about the reluctance of the NYSDOL to facilitate the hiring of domestic workers, in this case Puerto Ricans. Referring to growers' preferences for Jamaicans and Mexicans as "almost like a slave mentality," she declared that the rural reps avoided hiring Puerto Ricans because "they could walk off the job." After some aggressive back and forth with representatives from both the U.S. and New York Departments of Labor she continued, "You and I know the reason you bring in Jamaican and Mexican workers is control. They are abused; they cannot leave. Don't push it under the rug, you know more than all of us, you see it every day." Finally, the rural rep she was engaging responded, "It is my job to recruit. I beg them to recruit, but Puerto Ricans have told me, 'We're not apple pickers.' They choose the jobs. They go to North Carolina, Virginia, Georgia. They don't like the weather change [in New York]." Over the years I spent doing research in the field, I heard the claim that Puerto Ricans were avoided several times, and I even heard unconfirmed rumors that the NYSDOL was investigating why Puerto Ricans were not hired on the state's farms. More than one informant who was familiar with the NYSDOL practices confirmed that rural reps would "close job orders" if Puerto Ricans applied and then reopen them a few days later to avoid having to respond directly to those applicants.

Service providers and farmworker attorneys reported to me that blacklisting was sometimes used as a last resort to control hiring, and that this was done with the overt assistance of the NYSDOL. "If the employer doesn't get along with workers," explained a farmworker attorney, "the NYSDOL puts that worker on the list, on the back burner. They send the worker to an employer they don't have such a great relationship with. But these things are really hard to prove." Another reported a specific incident involving a farmworker client of hers who found a new job through a rural rep. She heard through another party that the new employer reprimanded the rep for the placement. The farmer did not want an employee with a history of instigating legal action against a previous employer.

Guest workers told me that their home countries kept blacklists as well. (Such practices are discussed in detail in the work of David Griffith.)[39] Although these techniques serve as an immediate form of labor discipline, they are not a direct response to growers' needs. The home country has a vested interest in sending quiescent workers, because it benefits from the guest worker programs in several ways. These include the creation of government jobs to administer the program, recouping a percentage of workers' wages as processing fees, and the provision of employment for the country's citizens. The relationship between the United States and the sending country is an unequal one, and, like the farmworker, the country is dependent upon the United States for the benefits of the program. If a sending country were to have a reputation for sending workers who complain and try to change the conditions of their labor, growers might discontinue requesting workers from that nation and the country's contract with the United States might be threatened.

Any understanding of the ethnic succession in today's labor market has to take account of long-established features of farm work in New York. Throughout the region's agricultural history, the work was arduous, the wages meager, and the employment precarious and usually seasonal. As a result, it captured the most vulnerable workers of all. While some farmhands served as apprentices, the vast majority had little expectation of job advancement and took on farm jobs out of sheer necessity.[40] They have been immigrants of various sorts, African Americans, and poor rural whites—all of whom were desperate for jobs during economic downturns, such as in the 1880s and the 1930s. As in the past, labor costs now drive growers' decisions about hiring, yet there is little public rhetoric within the agricultural industry that explicitly acknowledges this long-enduring fact. Instead, growers rely on well-polished mythologies to justify the low wages and poor benefits that they dole out to employees. These myths also extend to the poor working and living conditions that accompany agricultural work.

The vast majority of the farmers I interviewed took pains to communicate how it was difficult to recruit reliable labor and next to impossible to find local workers. This same complaint has been heard at different moments in the twentieth century, and in fact it was the main rationale for the creation of guest worker programs.[41] Today it is common in the industry to hear "No one else will do this work," or "I cannot find any local workers." The farm labor advocates I spoke to, however, argued that there is no shortage of labor. Rather, the problem lay in the substandard pay and work conditions, especially since farmers are not required to pay overtime rates. The former and current farmworkers I met—Latino and Jamaican—did hesitate to confirm that Latinos were cheaper and that workers' undocumented or guest worker status meant they could not challenge their employers. Each of these views can be accommodated by the dual labor market theory, which seeks to explain the segmentation of opportunities and wages.[42] But can this theory help us understand why international migration to the Hudson Valley increased so rapidly only at the end of the twentieth century?

The jobs on offer were not new jobs, as one of my respondents observed. "It is the same work, not the same people." First, we need to keep in mind that African Americans, particularly the poorest, were acutely disenfranchised, both politically and economically, for so long that they were effectively treated like noncitizens. Earlier farmworkers, like slaves before them, were also categorized as constitutionally inferior to whites and therefore incapable of taking on any but the most menial jobs.[43] In addition, many were acculturated into farm labor through family and community backgrounds, and, in the absence of advanced formal education and job skills, they had few other options. The departure of African Americans from agricultural labor can be attributed in part to the gains of the civil rights movement and legislation of the 1960s, when public employment opportunities opened up. One older farmer attested to the improved prospects for blacks, saying, "What happened to southern blacks was they got better jobs or learned to work the welfare system better. . . . They did not get aged out; they got more." From a macro perspective, black workers were then in a position to leave farm work—something most workers would do given the opportunity. Those who stayed on were more inclined to make demands on their employers. As Jim Schmidt, the former director of Farmworker Legal Services in New York, explained, "People started to see blacks as asking for more, or being uppity, which they never did before." In addition, it should be pointed out that wages for farmworkers in New York declined in the 1970s and 1980s,[44] and thus workers would have to make demands on farm owners not only to improve their situations, but even to maintain them. On the basis of this evidence, it is fair to conclude that dual labor market theory does not fully take into account the replacement of a domestic work force with a new immigrant one, nor does it help us understand the decline in wages that preceded their arrival.

Not surprisingly, the farm owners I interviewed were forthright about their concern over labor costs. One young grower who saw himself as part of the new green wave of agriculturalists reported that he had peers who wanted to expand but did not have the labor resources to do so because it was so difficult to find workers. Under those circumstances, he explained, "the dark door opens up," and farmers like him seek out undocumented workers. In his circle of alternative and organic farmers, this topic, he told me, was "something that gets talked about in hushed tones." In general, those who want to go green and produce boutique high-end produce and products were preoccupied with maintaining the public perception that they did not exploit workers. For more traditional farmers like Trask, hiring the undocumented was a more straightforward feature of twenty-first-century New York agriculture. Labor costs and labor stability were variables that determined whether they could grow their enterprises successfully. "I've increased my risk tenfold," Trask observed, saying, "I could fail, just like the next guy. I have a lot at stake"—stakes that included his investment in and responsibility to his workers. Musing about an acquisition that doubled his farmland, he confessed, "My debt load is the biggest I've ever seen. When I bought that farm I had sleepless nights. I still do."

Most farmers claimed that they feel constant pressure to reduce labor costs. They wanted to pay their workers more, they said, but they could not afford it. They insisted that they would be put out of business or would have to restructure or downsize their operations to accommodate the extra costs of overtime pay. The most prevalent argument used by growers and their allies to explain the perpetuation of the new immigrant work force is that the low-end sector of the secondary labor market suffers from a worker shortage. Growers complained persistently about this lack of labor supply, and temporary shortfalls no doubt occur. A grower may anticipate needing four workers based on experience, but when the harvest arrives she might suddenly find she needs two more. Supply and demand in agricultural production is not an exact science, but the challenge of finding new workers on short notice is not evidence of a chronic labor shortage, "Farmers and farm managers who recruit immigrant workers commonly view either a shortage of highly disciplined, reliable workers, or the absence of a *surplus* of workers, as a shortage of agricultural labor."[45] The reasons for the shortfall have everything to do with the low compensation offered for arduous and physically demanding work.

Employers' firsthand perspectives also shed light on the reasons for the demographic shift in the New York work force from blacks to Latinos. My interviews revealed a general disparagement of the former on grounds related to the following: age, drugs and alcohol, violence, race and ethnicity, and welfare use. In general, domestic employees were characterized as unstable and unreliable, or it was explained that they had "changed" over time. On the other hand, employers tended to praise new (undocumented) workers for their work ethic and loyalty, but also to express fear that they too might change over time. A desire to achieve the American dream was also regularly cited as a reason Latinos were such willing workers.

Growers' stories about drinking and rowdy behavior by former workers were also common. The degree to which this was regarded as a problem varied from one farm to another. As one farmer put it, talking about local black workers, "You could bring some people out of Hudson or wherever and they'd be here for a day, get some money for their drugs, and they'd be gone."

One Western New York service provider alluded to a crack epidemic on farms in the 1980s. She reported that drug use was directly responsible for accidents, such as falling out of trees and off ladders. In her thirty-year career with farmworkers, she divulged, "That was the only time I ever felt fearful about going into labor camps, because of drugs." Notwithstanding their occasional basis in fact, stereotypes of shiftless black workers, which go back to the plantation era,[46] were readily combustible fuel for explaining the preference for immigrants. Hard drinking is a feature of all-male labor crews in every work sector, especially among those who live within the confines of a remote labor camp. Race has nothing to do with it. One farmer reported that his Mexican workers drank "like fishes," but he thought nothing of it, so customary was this conduct among field hands. Griffith similarly describes how at least one Western New York grower in Wayne County attested to the changing character of his African American workers once Latinos began to arrive, recalling how some of the members of the African American crews were "rowdy and cantankerous, challenging Latino workers to fights in the labor camps at night. From one season to the next, more and more of the remaining workers on the African American crews became, in varying ways, more difficult. . . . Violence in the camps became common." The farmer blamed the longtime crew leader for recruiting increasingly violent workers. From this account, it appears that the farmer did not realize that the transition away from African American workers was a factor in the growing conflict. By contrast, his summary of the predicament implied that the farmer had little choice but to get rid of his African American workers for the safety of the Latinos.

Agricultural employers have long deployed ethnic stereotypes to hasten demographic transitions in the work force. Incoming or preferred workers are praised for their strong work ethic, while outgoing workers are castigated as lazy and overly demanding. Race-based characterizations are vehicles for employers' rationalizations about who will be good workers.[47] This kind of racial profiling, which is repeated whenever a new group is introduced, also intersects with employers' ceaseless search for quiescent workers to fill low-paying jobs.[48] Consider the advice offered in a prominent farm management textbook from the early twentieth century. The author classifies workers into groups largely according to race, ethnicity, and nationality in order to "indicate the laborer's more outstanding social or class traits, good and bad." Whites are designated as the best laborers but likely to go into business for themselves. Hobos and tramps are considered unstable troublemakers;

Italian and Portuguese immigrants are some of the best workers but not tolerant of strong discipline; blacks are good-natured and "exceptionally well developed physically" but are careless and constantly borrowing money. Mexicans are described as "peaceful, somewhat childish, rather lazy, unambitious," but very loyal. The book goes on to summarize the traits of Indian (Native American), Japanese, Hindu, and Chinese workers, a variety that reflects the many types of workers in California agriculture at the turn of the twentieth century.[49]

Management advice still profiles workers according to race and ethnicity, although today it is couched in terms of cultural differences.[50] For example, Cornell University's Department of Applied Economics and Management (part of the College of Agriculture and Life Sciences) sponsors conferences and research to help farm owners better understand and learn strategies for managing their Latino work force.[51] The discourse no longer relies on racial essentialism but instead appeals to cultural predilections, such as the aptitude for teamwork or attitudes toward authority, while advising on how to use praise or manage communication differences between Latinos and whites.[52]

Professional advice forums may have moved beyond racial essence, but the daily discourse of farmers has not. A white office employee on a fruit and vegetable farm described the racial characterizations she had heard from different farmers. "Apparently Puerto Ricans are really lazy," she started. Obviously embarrassed to have repeated the stereotype, she made it clear these were not her opinions:

> I hear what other farmers say. Mexicans are harder to deal with. They don't get along with the Guatemalans.... I hear that Jamaican workers are lazy and do a lot of drinking.... Puerto Ricans, I've heard concern about a bad attitude.... Americans are the worst. You never see them out in the fields. It wouldn't work, there would be too much attitude, they would be too lazy, they wouldn't make enough money. The work is too hard.

While stopping short of offering such comprehensive typologies themselves, my interviewees made specific comments relating to their own workers. One opinion I often heard was that black workers would rather collect welfare than work, and that the government makes welfare too attractive an option.

Comparison between types of workers was very common. A Dutchess County fruit and vegetable grower explained why the Mexicans were such good workers, saying, "You wouldn't get anybody else to work on a Sunday."

Pointing at me, he asked, "If you worked five or six days a week, you wouldn't want to come on Sunday as well, would you? No way. That's the difference between the Mexican and the black. For the Mexican, this type of work is the only work he knows."

Moreover, racial and ethnic characterizations shifted over time. When workers become empowered either through legal status, increased job opportunities, or some other factor, even the most idealized groups could become undesirable. One farmer told me that the downside of a Latino work force, for example, is that they can get "too Americanized" over time.

Ethnic and racial characterizations were offered not only by individual farmers but also by representatives of the agricultural industry in New York. For example, when the New York Farm Bureau sends materials to elected officials it commonly includes newspaper clippings. One such article, from the *Daily Messenger,* an upstate newspaper, was distributed to state legislators in an effort to rebut farmworker advocates' claims about poor conditions for the state's farmworkers. The article reiterated some of the more prevalent myths, and especially those that appeal to "ethnic" work traits. A dairy farmer is quoted as saying, "When the Mexicans came, it was like night and day.... These guys know how to work.... They know when they are supposed to be at work, they do a job, and they respect their coworkers.... They have a work ethic.... They don't cut corners."[53]

Ethnic profiling also commonly extends beyond the workplace to assessments of workers' tolerance of substandard living conditions. One farmer articulated a common myth about Mexican workers, whom he housed on his farm, saying, "They are not afraid of tight quarters. That gets more bad press for the farmers than anything. The farmer doesn't have control about how many will be in the house. The workers don't mind that; they like that." He went on to explain Latino workers' suitability for farm work, arguing that their "unique culture" meant they did not mind doing "work that will cause them to sweat or get dirty." He also pointed out, "They take pride in this work, that this is their farm. They know when certain chores have to be done and they get after it." Also quite common was the perception that different ethnic groups do not get along. This was a particularly serviceable explanation for why the ethnic shift took place so quickly or so comprehensively on some farms. A Jamaican apple picker, who assured me that Mexicans kept to themselves, also reported that farmers separate workers according to race, and that the language barrier prevented any real camaraderie. "They are working right beside you and you can't communicate," he said. Discrimination against blacks by Latino workers was widely reported. Ethnic pride also asserted itself in a competitive fashion. Many workers, for example, sung the praises of their own group and argued that Mexicans/Jamaicans/ Guatemalans "do the job better" or "are the best workers."

Growers projected onto workers a sense of pride in, or loyalty to, the farm itself. The farmer's quote in the previous paragraph also includes another layer, common to farms, small businesses, and domestic employment situations. When a farmer asserts that "this is their farm," he obscures the property relations that govern the balance of power between employer and worker. Allied with this misplaced loyalty is the characterization of workers that I heard from several growers and encountered in the local press: "My workers are part of the family."[54] A farm manager I spoke with described what the other side of being "like family" amounted to. He explained how he worked his way up the ranks on a farm; he was well paid and treated well. He was even invited to family events and was the pallbearer at a family funeral, but he "still wasn't family."

The willingness to work as many hours as requested combined with an unquestioning acceptance of labor discipline was bound to be interpreted as a strong work ethic, but it more likely represented farmworkers' desperate need for income, their will to sacrifice to support their families, and their fear of retribution for speaking out. One of the NYSDOL representatives I interviewed commented that he thought the Mexicans did have a superior work ethic and that they were seldom seen as idle, but he also acknowledged how difficult it was to separate that perception from their acute financial need. He said, "At five dollars an hour, they understand they have to put in the hours. If they could make more an hour, they would work less hours." He also conceded that the perception was racially loaded, saying, "Farmers say the Mexicans work better; it's a bit racial." He further explained that the industry had its share of racial segregation and discrimination on the part of both growers and workers, and even within the NYSDOL, but he took care to emphasize that it was somewhat subtle.

The increasing presence of immigrants in the labor market and the hiring decisions that sideline American workers should make it clear that the local food economy is not going to produce jobs that benefit existing community members. Moreover, the choice to move to a more vulnerable work force creates a new level of social inequality, which is not created by the immigrant farmworkers themselves, but rather by their employers and the industry at large. Growers and their interest groups did not create the conditions that limit farmworker power—they are rooted in a long history of the marginalized farmworker job category and found within specific moments of anti-immigration sentiment—but employers have shaped and managed these conditions for their own benefit. Business practices that depend on the exploitation of low-skilled workers commonly lean on racial discrimination in hiring processes.

In this chapter I have offered a comprehensive description of increased immigration in a particular regional industry and have analyzed the rationale of employers who facilitated ethnic succession in their work forces. Controlling workers and controlling hiring are intrinsically related. Through kin networks, recruiting and hiring guest workers, and finding employees through the Rural Employment Program of the NYSDOL, growers were able to determine the ethnic composition of their work force. Although larger political and economic factors have influenced blacks to leave farm work and undocumented Latinos to enter the sector, those factors do not fully explain the dramatic shift in New York's agricultural work force between the late 1980s and the early twenty-first century. As I have shown, growers, with the help of the state, actively controlled job access in order to favor undocumented Latinos and to limit access to workers who are citizens or legal permanent residents. They did so to secure a cheap and quiescent work force as a way of maximizing their profits. Moreover, the shift has been systematic and not sporadic or casual, with individual growers emulating others on a statewide scale. The power disparity that growers enjoy over their employees is continually reinforced by their ability to exploit the most vulnerable workers.

FIVE

Toward a Comprehensive Food Ethic

ALMOST EVERY WORKER I MET agreed to speak with me for this research. With few exceptions, they took the interview process very seriously. Cecilia, an apple packer from El Salvador who was employed on an Ulster County apple farm, leaned across the table and put her hand on mine; she made eye contact as she declared, "You must get our story out, for we cannot tell it." In part, to honor her request, I have tried to present her story and those of workers like her across the region. It was equally important for me to include in these pages the views of farmers to offer a more balanced perspective on the labor market and the challenges experienced on all sides. As I conclude this book, I return to the words of Cecilia and consider more deeply the implications of this research both for farmworkers who cannot tell their own story and for how their plight bears upon the quest for ethical eating.

As a locavore myself, I appreciate forms of agriculture that offer an alternative to the corporate industrial farms that dominate the traditional food system. Farm-to-table exchanges appear to take place in a self-contained, grassroots universe that lies beyond the orbit of corporate or government networks, and therefore in a realm that seems guaranteed to deliver virtuous outcomes. Indeed, I have learned quite a bit from the farmers I support about how my food is raised or grown, and I do feel a connection to them, as Michael Pollan and other writers have led me to expect. Yet, for the most part, "alternative" is used to describe a different means of offering high-quality farm products—in other words, an alternative market mechanism—as opposed to describing a truly different moral relationship to the process of agricultural production. What Marx called "the relations of production" have not been fully called into question in alternative agriculture. That is not to say that alternative agriculture does not have its virtues. It offers fresh, seasonal, diverse, and organic goods; it allows us to support smaller farmers instead of industrial agriculture; and it helps keep our food dollars local.

But if the revolution in agriculture that I and other locavores want is to fully embrace sustainable livelihoods as well as sustainable growing, then we need to develop a more comprehensive food ethic. We need to ask how our food choices might help make a better world.[1] In that spirit, this chapter builds on my earlier call for the inclusion of labor rights alongside environmental concerns and animal welfare in conversations and activism related to the food movement. It does so by investigating the strenuous but ultimately stymied efforts of advocates to improve the situations of New York farmworkers. The work of advocates illuminates how blocs within the grower community harness the power of the agrarian legacy to thwart reforms. In examining those advocacy efforts and explaining why they have not been more successful, I extract some policy lessons and recommendations for how farmworker justice might be advanced while also providing consumers with options for how to address their nascent labor concerns.

THEY DON'T EAT THE WORKERS

A few years ago I attended a seminar with a speaker who was a *charcutier*, someone who butchered and prepared meats on a small farm. He reported that in terms of animal welfare and food contamination, small farms actually had a worse record than larger ones. Outlining some of the problems inherent in the slaughter process, he noted that small farmers did not possess the high-tech equipment of industrial agriculture that guaranteed a quick death. During the question-and-answer period I asked him why he thought consumers seemed so concerned about environmental standards, pesticide use, and the treatment of animals but not the working conditions of farmhands. He responded drily, "They don't eat the workers." He went on to explain that, in his experience, his consumers' primary concern is with what they put in their bodies, and so the labor standards of farmworkers simply do not register as a priority. This was not the first time I heard this (though it had never been put so pithily), and it needs to be examined more thoroughly.

I am not going to offer here a definition of "sustainability"—a term that is subjective, contested, and susceptible to highly varied usage—but I will observe that those who use the term usually do so in the spirit of accountability not only to biotic health, but also to community justice. The concept is not usually applied to a purely individualistic pursuit of justice, health, and self-sustenance. Sustainability arguments derive from larger concerns: anxiety about resource depletion or the environmental impacts of our actions, dependence on an ever-limited supply of fossil foods to transport our foods, or a renewed emphasis on how consumers can play a role in promoting smaller-scale farming efforts that are more natural, humane, and community-oriented. It is within these more expansive contexts that we can talk about a food ethic as fully sustainable.

Yet if the foodies who are nurturing the alternative marketplace are motivated only by their own health and that of farm animals, what does that say about the recurrence of the term "sustainability" in so much food writing? Are we misinterpreting or misrepresenting consumer interest? Are a handful of idealistic food writers and consumers imagining and shaping this cause in an overblown way? Is the goal for locavores to feel more justified in their own food choices, or is it to foster deeper change in the structure of agricultural markets and food production? These kinds of questions call for further research about how recurrent concepts like "sustainability" and "local" are used as marketing tools as opposed to representing a new social movement in farming. Are they headed in the same direction as "all-natural," to serve as rhetorical labels for higher-priced commodities? More ominously, how does the marketing use of this language act as a salve for consumers' bad consciences while shielding them from the reality of poor labor conditions?

In New York, for example, the state Department of Agriculture and Markets offers grants to farmers to help them remarket their products as local to take advantage of consumer interest. In addition, while the CSA model was originally conceived as a project that would start at the community level and be driven by a core group of volunteers, now conventional farmers are creating their own CSAs as another channel for marketing directly to consumers, adding to a repertoire that includes stalls in farmers' markets and roadside produce stands. My point here is not to begrudge farmers an opportunity to create renewed interest in their products, for surely they need all the help they can get to go up against corporate industrial growers. Rather, I want to deepen our expectations of what local and alternative agriculture can and should be.

In an essay titled "Vote for the Dinner Party," Michael Pollan asks if the U.S. food movement deserves its moniker. Movements, he points out, demand change. "People like me throw the term around loosely," Pollan writes, "partly because we sense the gathering of such a force, and partly (to be honest) to help wish it into being by sheer dint of repetition."[2] As a way of realizing such a wished-for movement, he emphasizes the importance of winning political victories through state and federal legislators—a feat that is appreciably more difficult than persuading consumers to vote with their food dollars. This is necessary because government plays an extensive role in regulating agriculture at every level. The labor question deeply involves the state, but to date state labor regulations have exempted farmers from offering overtime pay or a day of rest and from being accountable to collective bargaining laws. Bringing the state back into the picture is unavoidable if food is to be a social movement, even if it runs the risk of debunking the belief that the social virtues of agrarianism can be achieved through goodwill and communitarian action alone.

To build on Pollan's argument, consumers should consider whether they are prepared to regard food as a common good. If agriculture is subject to state protections due to the importance of the industry in feeding the nation, then it certainly appears that the state has already recognized food as a common good. As such, ethical food should be part of a revived commons. Pollan posits that the farmers' market is the new public square and a place to express our "communitarian impulses."[3] But a commons only comes into being through public action. *New York Times* food writer and cookbook author Mark Bittman reasons that consumers should have a collective voice in agricultural processes, reminding us that the Homestead Act moved 10 percent of public land into the hands of private farmers and that federal dollars continue to provide the infrastructure that supports farms nationwide.[4] Yet, at least so far, the locavore version of the commons does not include those who raise, tend, and harvest what we put in our mouths. Pollan's call for state action is surely a noble one, but let us set it alongside the reality of what a campaign aimed at lawmakers actually looks and feels like. A closer look at a legislative fight for reform will illustrate the entrenched obstacles to activism and demonstrate more fully how the state functions to assist some agricultural interests and marginalize others.

NEW YORK'S JUSTICE FOR FARMWORKERS CAMPAIGN

In the summer of 2010, the Farmworkers Fair Labor Practices Act reached the floor of the New York State Senate and was narrowly voted down by a vote of thirty-one to twenty-eight.[5] The major provisions of the law included: 1) overtime pay at time-and-a-half for farmworkers after ten hours a day, sixty hours a week, and a sixth day of work, moving to fifty-five hours three years after passage; 2) a twenty-four-hour period of rest required each week, with the provision that workers could voluntarily refuse the day of rest and that the first eight hours would be paid at time-and-a-half, moving to double time after that; 3) collective bargaining protections for workers on the state's largest farms, with a twenty-one-day "cooling off" period to avoid a strike at harvest time, and the creation of an advisory committee to set rules for resolving labor disputes. Although a number of factors account for the bill's failure, many stories in the media described the potential passage of the bill in apocalyptic terms, suggesting that it would portend the end of agriculture in the state. This prediction was fed by the romantic agrarianism and agricultural exceptionalism that have influenced farm policy in the United States since its founding—ideals that have also fueled the alternative food movement's message about promoting small farms. The public appeal of these ideologies allowed key legislators to obstruct the bill's passage, even though it came closer to passing than it had in a decade of campaigning.

How did New York farmworker advocacy get to this juncture? Building on local activist efforts in various counties, coordinated state-level efforts picked up momentum in the early 1990s with the emergence of the Justice for Farmworkers (JFW) coalition. From the outset, the JFW pushed for worker empowerment, labor law equality, and remedies for worker grievances through a three-pronged strategy of farmworker organizing, legal casework, and legislative campaigning. Like food movement activists, the farmworker advocates sought structural as opposed to minor case-by-case changes. Previous efforts around the country throughout the twentieth century had achieved some of these goals for workers, but they had registered little success in changing the overriding structure of agricultural employment.[6]

In 1967, a tour of labor camps in Western New York by New York senators Robert Kennedy Jr. and Jacob Javits, as well as Rochester-based hearings on farmworker conditions, brought widespread media scrutiny to what they described as "appalling and disgraceful" living conditions.[7] A few years later, Cornell University researchers published an ethnography of sixteen migrant labor camps, describing in detail the daily life of resident workers. A second volume analyzed the farm labor system and painted a bleak portrait of the lives of farmworkers.[8] Around the same time, a controversy erupted about poor farmworker housing at the Cornell University–owned Cohn Farm, and public attention to this problem intensified when the university administration secretly demolished the labor camp. Wayne County Rural Ministry, the United Farm Workers, the Agribusiness Accountability Project (led by consumer advocate Ralph Nader), and students and faculty at Cornell all played a role in highlighting not only the poor housing, but also Cornell's reluctance to address the issue head-on. The result was the creation of the Agricultural Manpower Program, which later became the Cornell Migrant Program, to address the needs of migrant farmworkers through outreach, education, and research.

In 1987 Governor Mario Cuomo vetoed state legislation that would have exempted farmers from the state's pesticide notification laws. Heated debates that pitted the departments of labor and health against the organized interests of the agriculture industry ensued. A similar face-off between farmers and labor advocates also made its way into public discourse after Governor Cuomo's 1989 State of the State address, which called for the creation of a Cornell University Task Force on Agricultural Labor Relations to conduct a study on collective bargaining rights in New York agriculture. The task force, funded through the state department of labor, recommended that farmworkers be extended the same rights as other workers, particularly in regard to collective bargaining protections.[9]

Further adding momentum to the state's farmworker advocacy was a 1992 campaign, led by Hudson Valley advocates, against sanitation code violations in farm labor camps. State senate hearings on farmworkers were held in 1994, and the corresponding 1995 senate report also called for an extension of rights to farmworkers.[10] As advocates hoped, moral outrage about farmworkers' working and living conditions in turn generated support for the campaign. With individuals and groups throughout the state primed to help, advocates mobilized this new wave of supporters to visit, write, and call their representatives in Albany. Encouraged by the response of a few state politicians, the JFW began to organize group visits to lawmakers, and, in 1995, the coalition initiated an annual public rally at the state's capitol. As a result of this pressure, New York State law changed to benefit farm laborers in 1996, 1998, and 1999. The JFW claimed victory.

Emboldened by those successes, the advocates refocused their efforts in 2000 on an omnibus bill—the Farmworkers Fair Labor Practices Act—aimed at removing all the labor law exceptions that benefited agriculture, including provisions for overtime pay, a day of rest, and collective bargaining protections. Protests, rallies, vigils, fiestas, fundraising dinners, and a robust email network of communication between advocates and their supporters, both in Albany and in the general public, helped maintain the momentum of the campaign.

The choice to focus on state mechanisms instead of relying on the momentum of farmworker organizing to create change stemmed largely from the difficulty of organizing. Across the United States, farmworker justice advocates have a long record of successfully targeting large-scale agriculture and/or corporate food brands. César Chávez and the United Farm Workers' (UFW) successful table grape boycott, which won improvements for California's agricultural workers in the 1960s and 1970s, is enshrined in the U.S. public memory. In the late 1980s, the Farm Labor Organizing Committee (FLOC) in the Upper Midwest confronted an industry dominated by five corporate owners, including Campbell Soup and Heinz, to win union contracts for workers, and, in the 2000s, FLOC successfully organized foreign guest workers in North Carolina. Since 2005, the Coalition of Immokalee Workers in Florida won agreements with eleven major food companies—Taco Bell, McDonald's, Burger King, Subway, Chipotle, Whole Foods, Trader Joe's, and the food industry service providers Bon Appetit Management Company, Compass Group, Aramark, and Sodexo—to increase wages for the state's tomato pickers. The achievements of these campaigns were largely due to national consumer boycotts after the model of the UFW grape campaign. They demonstrated how campaigns can organize consumer action around popular brand names. In New York State, however, the available targets are by and large neither the corporate vegetable buyers like Burger King, nor the industrial "sweatshops in the fields" of the largest commodity farms.

To the degree to which the brand-targeting campaigns relied on organizing farmworkers, this was a daunting task, even in states such as California that had vast single-location work forces.[11] In the Hudson Valley, and in New York State in general, it is extremely difficult to maintain consistent access to large numbers of farmworkers. Not only are organizers impeded by the small size of the work force on any given farm and the distance between farms, but also workers' relative immobility due to a lack of transportation prevents significant camaraderie much less solidarity from developing among workers at different farms. Moreover, a seasonal or migrant work force may be on a farm for only part of the year. And, as I experienced myself when I accompanied farmworker organizers, they spent many long hours on the road. At their destinations, they invariably found that workers were not always at home, and appointments were often forgotten or postponed. Much of the organizer's work needs to be carried out during the evening hours, but this is also when workers prepare their meals and relax.

Even when organizers can achieve consistent access to workers, convincing the workers to strive for change is even more challenging. Farmworkers are unwilling to risk what they have, particularly since most of them made huge sacrifices just to reach the United States. As one organizer acknowledged, many workers are afraid even to meet with outreach workers from local service organizations. "They would like to change their reality," he observed, "but they don't want to risk their livelihoods." Another put it this way: "They stay quiet even though they are suffering." Both affirmed that undocumented workers have little expectation of their circumstances changing, and they have no reason to believe that they have the power to effect change. Lawyers and paralegals both spoke to me about farm laborers' pressure from coworkers to ostracize anyone who broke ranks to lodge a complaint. In addition, organizers regularly heard from workers that farmers explicitly warned them not to associate in any way with organizers. In general, I had the impression that most workers, fearful of employer repercussions, did not want their concerns raised until the end of the season, after they had left the farm, and legal service providers and a Department of Labor representative told me this as well.

Confronted with such formidable challenges to farmworker organizing, advocates opted to focus on the passage of the Farmworkers Fair Labor Practices Act in the state legislature. Although members of the state assembly were generally sympathetic to the issue, the advocates faced a much greater challenge trying to win over the Republican majority in the state senate. In 1995, the assembly had passed a farm labor bill that supported collective bargaining protections, but similar legislation was not even acted on in the Senate Labor Committee for more than a decade. Why did these efforts fare better in the lower chamber? First, proportional representation in the assembly meant that farmers (who generally hailed from rural, Republican districts) were much less influential in the assembly than in the upper chamber. Second, the assembly had a long-standing Democratic majority, making it more likely to pass labor legislation, a traditional policy interest of urban liberals. Finally, the top-down nature of Albany politics made it almost impossible for any bill to get floor time, let alone move to a vote, if the leadership did not approve it in advance.[12]

Backed by a growing contingent of concerned New Yorkers, increasing public awareness of labor conditions on farms, and a handful of powerful legislative allies, the advocates took responsibility for the passage of three pro-farmworker measures in the 1990s.[13] The first was a 1996 bill requiring employers to provide drinking water for all farmworkers; previously the law had only covered farms with five or more workers. In 1998, the legislature passed a bill requiring portable toilets and hand-washing facilities in the fields (or transportation to sanitation facilities) for all farmworkers; earlier legislation only covered farms where the workers numbered eleven or more.[14] The third vote for which the JFW claimed credit was the 1999 bill that raised the agricultural minimum wage and tied it to the state's minimum wage—an increase of 21 percent at the time. These successes not only built momentum and attracted additional supporters, but they also convinced advocates that more comprehensive reforms might be possible.

Legislators typically tend to reward those who provide electoral support. Industry interest groups effectively rally supporters because proposed regulatory legislation affects not only the industry as a whole, but also individual businesspeople. This generally makes business owners easier to organize since they have a direct interest. Consumers and those in favor of industry regulation, on the other hand, are not usually directly affected by proposed legislation. In short, this amounts to a collective action problem. Industry is highly motivated to organize to support or fight policy proposals, but consumers usually are not. In turn, industry exerts a considerable influence on legislators' decisions about regulation.[15] When fully mobilized, the agricultural industry is capable of shaping legislators' opposition to bills that threaten to alter the balance of power in the workplace.

Grower power is also underscored by the close ratio of farmers to laborers. In New York State, according to the 2007 Census of Agriculture, there were 36,000 farmers and 60,000 workers.[16] In very few industries is the ratio of employers to employees so small. This gives owners a matchless advantage over their employees when it comes to contacting and lobbying their elected representatives, not to mention influencing their neighbors, friends, and acquaintances on policy issues. Since farmworkers have so little political clout and most rural legislators have little interest in their mostly nonvoting and often seasonal farm labor constituents, the workers' advocates and allies are an especially critical voice. After many years spent petitioning Albany lawmakers, Richard Witt, the head of Rural and Migrant Ministry, was in a position to confirm what an aide to a prominent state senator once told him: "The Senator doesn't care about farmworkers—they have no political voice."

The JFW's legislative successes in the late 1990s and its growing public support were met with increasing opposition from farmers and their organizations, especially the New York Farm Bureau.[17] This opposition courted legislators and offered their own educational briefs on farm labor issues. In addition to contesting policy, members of the agricultural industry actively tried to sabotage the JFW's campaign by questioning the legitimacy, legality, and ethics of the organizations and individuals involved. These efforts were aimed at generating negative press for farmworker advocates or at terminating or limiting their funding by pressuring supporters and funders of their nonprofit organizations. In addition, farmers themselves reported the advocacy organizations to the New York State Lobby Commission and the New York State Attorney General for violating lobbying regulations. For organizations that received state or federal funding, farmers barraged state legislators and members of Congress with an array of complaints designed to trigger investigations.[18] Farm owners also targeted individuals (including this author) and defamed them in an attempt to destroy their reputations, especially with funders.

The backlash has ranged in type and significance. Violence underscored the resentment of farmworker efforts in many states in the early and mid-twentieth century, and this has also been apparent in New York. In a 1989 lawsuit farmers were accused of violence against legal services representatives who were on the farm to meet with workers; the farmers were accused of smashing their car widows, slashing their tires, and pinning lawyers inside their cars with farm vehicles. In addition, in a 1994 federal class-action lawsuit, the same farmers were accused of beating a worker. Both cases were decided in favor of the plaintiffs.[19] In 1998, the primary passenger van of the nonprofit that was trying to organize the state's farmworkers (Centro Independiente de Trabajadores Agrícolas/Independent Farmworker Center) was vandalized so badly that it had to be replaced. During the 2003 JFW ten-day march, the main vehicle carrying supplies was found with bullet holes through a window, shot right through a "Justice for Farmworkers" sign that was hung in a window. (Police investigations of these incidents yielded nothing.) Advocates have also described being aggressively confronted by farmers when visiting labor camps, sometimes with shotguns, heavy farm vehicles, or police officers. Such backlash against farmworker advocates is part of a long trend in the United States.[20] As organizing became more widespread starting in the 1930s, growers and their associations used physical aggression against workers and union organizers.[21] In the political sphere, they have used their concentrated power to influence federal policies and crush the initiatives of farm labor pressure groups.[22] In addition, growers prevented the federal U.S. Department of Labor from improving farm labor conditions that were part of the comprehensive New Deal effort to establish fair labor standards for all workers.[23] The agricultural industry has also actively prevented research from being conducted on the field work force at state agricultural colleges.[24]

One final example of backlash relevant to the New York case was the dismantling of the Cornell Migrant Program (CMP), a move that illustrates the behind-the-scenes power of industry interest groups. The CMP operated for more than thirty years as a university and cooperative extension program designed to assist the farm work force and increase awareness of their situations through research, education, and outreach. In May 2004, after an internal review and concerted pressure from farmers and agricultural interest groups, Cornell University administrators removed the program's longstanding executive director and moved the program into the College of Agriculture and Life Sciences, the traditional training ground and research arm for farmers in the university.[25] In the review report, there was no specific critique of the executive director's record of management, the programming he oversaw, or the CMP's approved strategic plan. The executive director's removal, along with other proposed changes, compelled all twelve staff members to leave the program,[26] and the major outside funder (the New York State Department of Education) to withdraw its support of more than $600,000.

The JFW advocates were convinced that industry representatives worked closely with Cornell deans to exert their influence over the program review. Uniquely, the CMP was not designed to serve growers; it was one small program for farmworkers established to counterbalance the numerous programs that support growers, including several at the Cornell Cooperative Extension, at the New York State Department of Agriculture and Markets, and at the College of Agriculture and Life Sciences within Cornell itself. In the thirty years the program was in operation, farmers were never considered stakeholders. Yet, in the review report, six of the eight stakeholder groups were afforded only a small paragraph each, while two of them—farmers and farmer associations— were given a full two pages, about two-thirds of the total space in that section.

My own review of extensive interviews conducted for an oral history of the CMP confirmed that farmers and others agreed that farmers and industry representatives were responsible for bringing down the program.[27] It is difficult to determine to what extent the industry influenced Cornell to change the program, but farmers themselves believed it to be extensive, and this empowered those engaged in backlash against farmworker advocates. The CMP change—it was resurrected as the Cornell Farmworker Program with a new mission and leadership—left a significant hole in the resource landscape, as it had played a role in coordinating relationships among farmworker service providers and advocates in the state.

The reactions by the nonprofits to such backlash ranged from frustration to anger, fear, and even a sense of validation that the advocates were finally getting somewhere. As proof that the backlash was relatively successful, several service-providing agencies that had been active in the JFW stepped back from the legislative campaign, and others reported stepped-up pressure on their already strained resources. For example, both the Rural and Migrant Ministry and Farmworker Legal Services were in danger of closing their doors entirely due to new financial constraints. The budget of the Centro Independiente de Trabajadores Agrícolas from 2006 onward allowed for only one salary, and so it quietly closed its doors in 2010.

While the backlash was picking up pace, the more public approach to influencing farmworker legislation was to try to strategically frame the debate around the omnibus bill itself. Both sides vied to "educate" and train the public in how to think about farm laborers' conditions and rights. Advocates defined the public debate as a matter of morality and human rights. The charge of moral injustice was backed up by the religious profile of the Rural Migrant Ministry, the main nonprofit within the JFW.[28] State senator Bill Perkins used stronger language yet by tying farmworkers' conditions to the "country's original sin of slavery" in his promotion of a 2009 state senate resolution honoring the state's farmworkers and calling for improved labor rights. Since it was merely symbolic legislation, it was considered noncontroversial and passed with a short roll call vote.[29] Kerry Kennedy went further in a Thanksgiving Day op-ed in the *New York Daily News* that began, "There's near-slavery in New York."[30]

The New York Farm Bureau's chief lobbyist countered that the effort of advocates to frame labor laws on moral grounds was inappropriate.[31] The proper perspective, according to the bureau, was an economic one, and not only for farmers, but also for workers. Accordingly, the organization estimated that mandating overtime pay would put farms out of business. Alternatively, farmers would either hire more workers to prevent overtime hours from accruing or else downsize their operations. Either option, the New York Farm Bureau argued, would be detrimental for workers. Moreover, the farmer organization argued against the collective bargaining measure by expressing concern that union dues would reduce workers' pay.

Deftly manipulating the locavore angle, New York Farm Bureau president Dean Norton wrote a plea titled "Don't Turn Farms into Factories" in which he argued that the passage of the Farmworkers Fair Labor Practices Act would decimate the "personal touch" of farming and that the state's diverse crops would become commodities like corn, oats, and soybeans. He argued that local food, in particular, was under threat.[32] Norton and other growers also promoted the idea that farmers are also farmworkers, positing that they enjoy fewer protections than their hired hands.[33] The typical New York farmer is often presented as someone who works longer hours and earns lower wages than his employees, collects food stamps for his family, and takes full responsibility for the risks involved in the enterprise. Few people doubt the hard work and sacrifice of the state's farmers, but to conflate the status of farmers with that of their employees and to actively promote the idea that farm owners are somehow more vulnerable than their workers obscures the reality of farmworkers' situations.

After ten years of effort, the Farmworkers Fair Labor Practices Act was finally heard on the floor of the state senate, thanks to the new Democratic majority and the maneuverings of Senator Pedro Espada and was voted down by a margin of three votes.[34] The floor debate, on August 3, 2010, lasted less than two hours, but it resonated with all the agrarian idealism, hardened dogma, and ingrained public perceptions I have described in the pages of this book. Senator Eric Schneiderman made a plea for correcting historic wrongs that were rooted in racism. On the other side, Senator Dale Volker labeled the bill a "death knell" for agriculture, a phrase repeated by two of his Republican colleagues in their speeches. In one rhetorical set piece after another, the farmers' economic plight and their right to maintain a work force at labor standards lower than those in other industries were pitted against moral arguments about the universality of human rights. Though some errant souls crossed the rural-urban divide, the state's enduring regional antagonisms (rural/urban) established the framework for the debate.

In the final analysis, the bill failed not on the strength of the arguments but because of a lack of political support for its passage. The morality and the economics of the bill were debated, but in the end they mattered little in the face of the balance of political power and possibly the hopes of key legislators to assure their reelection. Nonetheless, the rhetoric was key in influencing public opinion and enlisting support from reporters and op-ed columnists, community groups, allies, and other interested parties, all of whom play a role in putting pressure on politicians. In time, a mobilized public can and will shift the power dynamics in favor of some policy proposals. This bill's time may yet come if and when that happens.

To better understand this point, consider that during the same legislative session the Domestic Workers Bill of Rights, the first of its kind in the nation, was passed after a six-year effort.[35] Domestic workers and agricultural workers shared the ignominy of being excluded from the New Deal labor protections that provided overtime pay, a day of rest, and collective bargaining protections. The Domestic Workers Bill of Rights legislated a right to overtime pay after forty hours of work or forty-four hours for those who live in employers' homes, a twenty-four-hour day of rest or overtime pay if the employee chooses to work, and three paid days of rest a year. In addition the New York State Department of Labor was charged with conducting a feasibility study for collective bargaining. The protections granted to domestic workers essentially mirrored those sought for farmworkers. How can we explain that the same legislators dealing with two such similar bills passed one and not the other? The best way to understand the difference is that those who would be most affected, the employers, were of very different kinds. Farmers coordinated with strong business interest groups and enjoyed significant lobbying influence in Albany, while the employees of domestic workers were neither organized nor well represented (as employers).

The main themes examined in this book are also critical for understanding the failure of the legislation, because there are multiple, interlocking factors that stand in the way of improving farmworkers' situations. In the lead-up to the debate about the bill farmers appealed to the cultural cachet of agrarianism to influence the public and politicians in favor of their interests, and they were quick to utilize locavore rhetoric to cast themselves as victims of the corporate industrial food system. Over the decades, they have succeeded in cultivating the sympathy of the public for their plight as endangered farmers. Food writers have embellished the narrative about the wholesomeness of regional farms, highlighting and praising individual farmers and depicting local farms as the virtuous alternative to corporate industrial agriculture. In turn, consumers were given every reason to side with, and support, the intrepid producers who worked to bring fruits and vegetables to the public.

In addition, the historical and structural factors that have marginalized farmworkers have made it very difficult for them not only to seek redress for their grievances, but even to voice them. This has been further exacerbated by the influx of undocumented agricultural workers into the state in the past few decades. Moreover, workers' price of proximity meant that they struggled to maximize their opportunities within the context of paternalistic labor relations. State remedies were far from their minds. These workers not only lacked political power, but their situations were largely concealed from consumers, making it difficult for them to garner any public sympathy, let alone to compete for the affection lavished on their employers.

GOVERNMENT REMEDIES

In the course of the decade that I conducted my research, the public profile of the state's farmworkers was raised significantly. A decade ago there were very few opportunities for the average New Yorker to learn about labor conditions on farms. Since then, however, researchers, journalists, legislators, and advocates have used public channels to show how current farm labor relations are a product of long-standing patterns of inequity. Today the JFW has garnered renewed support from organizations such as the New York Civil Liberties Union, the Robert F. Kennedy Center for Justice and Human Rights, and New York University Law School's Immigrant Rights Clinic. Yet key legislative changes remain elusive.

Why is government action so important? Remember that the current law allows workers to log seventy- to ninety-hour workweeks at minimum wage (since overtime pay is not required). The law also does not mandate a day of rest, and workers can get fired simply for asking for a raise. Because of the price of proximity on these regional farms, paternalistic practices do allow individual workers to secure better pay or benefits, but such rewards are not codified in the law. As a result, workers tend to accept the conditions of their work in the hope of favorable treatment from the boss. The changes proposed in the omnibus bill—overtime pay, a day of rest or overtime pay if the farmhand does work that day, and collective bargaining protections—would all go a long way to changing the structural factors that marginalize workers. Eleven states have collective bargaining protections for agricultural workers.

Four states in addition to Washington, D.C., and Puerto Rico, have overtime pay provisions.

As for the feasibility of these alterations in the law, keep in mind that Hudson Valley farmers have faced myriad challenges over the past two centuries. Some have gone out of business, but others adapted to the changes, and the region still has a vibrant agricultural sector. Policy makers want to please all sides with legislation, but we need to understand if a bill such as the Farmworkers Fair Labor Practices Act were to pass, growers would need to direct more funds to their workers' pay. This directly hits their wallets, but the change would not be qualitatively different from other regulatory measures that ethical food advocates have campaigned for. Pesticide regulations that determine what chemicals can be used, how much, how often, and what sort of notification farmers need to give generated expenses for those who grow food. The "humane" treatment of animals—increasing the size of their pens, improving the quality of their feed, creating an environment where they can be free range or roaming, and improving the technology used in slaughter—was also accepted as a cost, however burdensome, of doing business in a world of shifting ethical norms.

We must not forget that the very low wages offered to farmworkers, the lack of overtime pay, and the lack of collective bargaining have all served to massively subsidize farming in this country.[36] Some will bristle at this analogy, but compare the situation to slavery. Just as the criticisms of the Farmworkers Fair Labor Practices Act by the New York Farm Bureau hone in on the economic cost to farm owners, a similar argument was once presented for maintaining slavery in the United States.

In the years I conducted this research I noticed a distinct change in the attitudes of farmers regarding the proposed legislation. In 2003 they used apocalyptic language—"it's going to kill me"—to describe the anticipated impact of required overtime pay. Six years later, they were trying to imagine ways of mitigating their costs if the bill passed. Though they may not have been aware of it, this psychological shift had occurred many times before in the record of Hudson Valley agriculture. Adaptation is one of the great legacies of farming in the region. In discussing how they might alter their practices, two practical questions rose to the fore: how could farmers afford to pay overtime, and how could they assure their workers the same number of hours and the same total pay as in previous years? The growers were concerned that they would have to hire more workers, which would mean fewer hours for everyone, or else they would need to cut back on their operations, which would have the same effect. To avoid these problems, the growers imagined implementing two strategies: 1) charge for services that were previously free, and 2) restructure hourly wage rates to accommodate the overtime provision.

Most of the workers I interviewed had either free or very inexpensive housing, many had free electricity and heating, and others had free satellite television service. Farmers who provide such services either for free or below what is allowable could modify their practices to make up for the overtime they expected to pay. Of course, farmers who are already charging the maximum for housing would not have this option. The other option is for farmers to lower workers' base hourly pay to compensate for the fact they would be paying overtime. For those paying above the minimum wage, there is some flexibility to deal with overtime regulations. Agricultural census data for New York puts the average hourly farmworker wage at above $10.[37] So, for example, if a worker logs eighty hours a week at $10 an hour, that would be $800 a week without overtime pay. If overtime pay were to start after fifty-five hours (as in the proposed legislation), then the employer could lower the base wage to $8.65 so that the first fifty-five hours would be paid at $8.65 an hour ($475.75) and the next twenty-five would be paid at time and a half, or $12.98 an hour ($324.50), for a total of $800.25. This scenario also works if overtime pay were required after forty hours. It would take some accounting acumen to estimate a new base rate for the season since workers' hours vary from week to week, and these calculations do not include other pay-related costs such as Social Security and disability deductions, but it could be a creative and smart way for growers to limit their costs. The farmers who would be most hurt would be those who pay minimum wage or very close to it, those already charging the maximum for housing, and those who require their employees to work an unusually large number of hours.

Farmers present themselves as "price takers, not price makers," meaning they have little opportunity to influence the retail cost of their goods and must accept what is offered. But when local growers offer their produce at farm stands, farmers' markets, or CSAs, they not only cut out the middleman who profits when goods are sold wholesale, but they also place themselves in a position to set their own prices. Consumers expect local foods to cost more, and they are generally prepared to pay. How much more would they pay for fair labor standards? The Coalition of Immokalee Workers fought for a penny more for a pound for tomatoes with the goal of increasing workers' wage rates. The resulting increase in the retail price would be negligible. In the locavore market, the increase might register more, yet the impact on the workers' yearly income could be significant. Labor economist Philip Martin has shown that a 40 percent increase in farmworker wages could be covered by a 3.6 percent increase in the cost of produce, or $15 a year for a typical U.S. household.[38] Let's put this in context. The average U.S. consumer spends less than 10 percent of her income on food. And of the food brought home, 15 percent is discarded.[39]

Another progressive and proactive strategy farmers could adopt would be to embrace the legally mandated rise in labor standards as a selling point. They would be able to promote themselves and their produce on the basis of these upgraded standards in much the same way as they do around the minimal use of pesticides and the humane treatment of animals. In this way they could speak directly to local food enthusiasts who are prepared to embrace a comprehensive food ethic. One such farm already serves as a model. The California-based Swanton Berry Farm advertises itself and its product as labor-friendly. The farmers take pride that their workers are represented by the UFW and receive health insurance and a pension. Its website points out that the farm accomplishes this on a "razor-thin" margin of 3 percent profit, or 18 cents on a $3 box of strawberries.[40]

Readers might consider these proposals and ask why such a law is important if workers make the same under a new law with overtime as they made without it. The significance of a government remedy is that it institutes formal rules instead of flexible or informal ones, an especially important distinction for the undocumented. With legislation, the structure of the workers' occupations will be altered to benefit them. The proposals are designed to suggest how to mitigate the initial burden on farmers. Once the system is in place, there is room for workers to get pay raises over time according to a pay system that resembles that of most other employees in the general work force, including domestic workers in the state.

LOCAVORE RESPONSE

Finally, what role can consumers play in improving the situation for farmhands on regional farms? Let us not imagine that the embrace of the farmworker cause by the food movement will resolve the whole range of inequities. However, this is an influential group with considerable sway over public opinion, and so any amount of added scrutiny and awareness would go a long way toward promoting farmworker interests.

First, educate yourself. Consumers can learn more about farmworkers in New York, the Hudson Valley, and other regional settings by doing some elementary research. The Internet of course offers access to many reports and articles. National organizations such as Human Rights Watch, Oxfam, and the Southern Poverty Law Center have all written extensive reports on U.S. agricultural workers. More difficult to find is research and resources on local workers. Many states have campaigns similar to the Justice for Farmworkers Campaign in New York with information on local agriculture. Other organizations such as Duke University's Student Action with Farmworkers and the D.C.-based Farmworker Justice have excellent resources. Books like *Tomatoland,* by Barry Estabrook, and *With These Hands,* by Daniel Rothenberg, both based on interviews with farmworkers, offer a wealth of insight into their predicaments. Delegations to meet with workers offer a more hands-on opportunity. For example, the Coalition of Immokalee Workers hosted ten national food activists on a tour of Florida tomato farms, and New York's Rural and Migrant Ministry has similarly brought allies to meet with workers to hear their stories firsthand. Always be cautious of farmer involvement on such visits. If the employer makes the introduction to workers, then you will be seen as an ally of the farmer, and workers will likely tailor their stories to be more palatable to their employer. Remember not to fall into the local trap; do not assume that a "small farm" or a "family farm" is necessarily a wholesome alternative to corporate industrial farms when it comes to the treatment of workers.

Second, ask questions about farmworkers. In the mid-1990s, it would have been very unusual to ask your waiter if the chicken special was organic and local, but today it is common. The same can happen in regard to labor conditions. The farmers I interviewed told me they often field questions from the public and journalists about details of the production process or about the efficacy of pest management practices, but rarely does anyone ask about the conditions of their workers. Consumer demands drive production decisions, and asking questions could promote more transparency about the workers' labor conditions, raise the profile of the workers, and possibly lead to better pay and conditions, much in the same way that the alternative food movement has improved environmental conditions and animal justice. Consumers, however, must educate themselves to understand farmers' responses and ask blunt questions about whether employees earn overtime pay, how many hours they work, and whether customers can speak directly with employees when the farmer is not around. Farmers need to hear directly from consumers that labor concerns matter to them.

Third, demand reporting on farmworkers. One of the barriers to adequate representation of agricultural laborers in food writing, policy reports, and media accounts is the dearth of data, profiles, and even recognition of the workers' situations. Farm owners tend to protect their labor force, particularly those who are undocumented, and themselves from negative scrutiny from the media, law enforcement, and neighbors. Although such paternalism may stem from good intentions, it further relegates farmworkers to invisibility. Scholars, consumers, and food activists can play a role in demanding more attention to labor concerns. For example, many farmers' markets have booths stocked with informational materials. None of those I have visited had anything on hand about the labor force, which I suspect is generally the case. Ask for brochures about the laws that pertain to them, worker demographics, human interest stories, or farmworker-developed recipes. Similarly, the media, including community newspapers, which regularly feature farmer- and food-related stories, could be asked to promote a more holistic picture of the production process and more fellow feeling toward farmworkers. Writers could profile laborers in the same way they feature individual farmers, with reference to their history, family, personality, and dreams. A little attention would go a long way. For a previous generation of shoppers, the UFW grape boycott raised the profile of this labor market and put laborers on the national stage. Today's emphasis on celebrating food and scrutinizing its patterns of production could easily serve a similar purpose by throwing some light on the labor behind the produce. Consumers can further help by using social media to highlight such stories and using their own blogs to explore the issues.

Fourth, consider farm labor policy proposals. The profile of the food movement has moved beyond nonprofit policy centers into the vast realm of social justice committees associated with places of worship and university and high school campuses. Moreover, new coalitions, particularly between environmental activists and foodies, are on the rise, and there has been a rapid increase in university food studies courses across disciplines. This combination of nonprofit and grassroots organizations and students is an important network for influencing food policy. However, we must be aware of the participation of farmers on the boards of food-centered nonprofits and their role in shaping policy priorities. We cannot expect farmers to advocate for their workers. In 2009, for example, the Northeast Organic Farming Association—an alternative voice to the New York Farm Bureau, and a significant influence on the state's alternative food movement—shifted gears to oppose the Farmworkers Fair Labor Practices Act. Food activists must be willing to challenge farmers.

Fifth, be wary of getting lured away from labor policy. Performing charitable acts is extremely gratifying, and so many farmworkers are in such desperate poverty that donations are usually very much appreciated. Structural change, however, requires altering the law. Tell legislators how you feel; join a campaign, sign a petition, or attend a lobby day in your state's capitol. Immigration reform on its own will do little to change the structure of these jobs. Farmer organizations in New York have long tried to deflect labor activists' efforts to improve work conditions by calling instead for a focus on immigration reform and immigrants' rights. The farmers' representatives argue that if the workers were in the country legally they would be better protected in every regard. Although that might be true, it still leaves farm laborers without the labor protections afforded to most other workers. "Fair trade" certifications can include labor concerns, but one major problem with such certifications is the participation of farmers in shaping them. Although such efforts, like those related to offshore apparel production or to fair trade coffee, can be used to start a conversation about labor concerns, they do not offer the consistent and wholesale structural change that legislative action can deliver.

Sixth, support local farms to build a food movement that incorporates workers. Buy local! The more vibrant we can make regional agriculture, the more prosperous our farmers will be and the better positioned they will be to pass on their profits to their work force. Particularly in affluent urban areas, where food dollars are plentiful, consumers can have a critical influence on building up the industry. But tell farmers what you want your purchase to support, much in the same way that consumers do in the instance of pesticide practices and animal treatment. Make sure that petitioning on behalf of workers is not expressed as an attack on farmers. Instead, transfer the responsibility to your choice as a consumer; explain your food ethic and how it demands fair labor standards to be observed in the products you buy. In the last ten years we have seen an explosion of interest in farming by a younger generation that has developed in parallel to the expansion of the marketing of local food. Look at this group as an opportunity; they are experimenting with different business models, and you can propose they put sustainable livelihoods at the center.

Think about the many benefits to be accrued from restructuring agriculture around sustainable jobs and how improved labor practices can be promoted as a selling point. Just as Whole Foods has a point system to indicate how the animals that become our meat are treated, so, too, smaller farmers could advertise the benefits they offer their workers to explain food costs to consumers. A Hudson Valley or New York food culture could be developed as a distinct regional alternative to the industrial exploitation of farmhands. I have seen proposals for boycotting farms or outing those known to have poor practices. On the corporate industrial scale, this strategy has had some success, but my opinion is that when it comes to smaller farms, the carrot would work better than the stick. Bear in mind that the changes proposed will be regarded as quite radical, and that growers need some time to adjust.

. . .

In describing the challenges faced by those who might respond to the cause of farmworker justice, I have emphasized that the farmers themselves have the most to lose. This has always been the case. The chronic economic insecurity at the heart of their industry dates back to the era of tenant farming, even as it has been punctuated by short periods of stability and steady profit. The profits of most small farms today are directly predicated on workers' low wages and lack of collective bargaining, and so farm owners feel they have no choice but to fight against the extension of labor protections. Their main justification for maintaining current labor relations rests on saving farming in the state, with the corollary that improved wages and work conditions would put them out of business. I am in sympathy with the constraints on farmers, and I have highlighted how smaller farmers and their workers are both victims of the corporate industrial food system. Even among my grower interviewees, however, at least two were quite open about why overtime pay and other provisions were needed to help workers overcome exploitation on Hudson Valley farms. They intimately understood the implications for their businesses, but they felt strongly that such provisions were a matter of basic justice. In addition, several others described the system as unfair to workers and suited only for the most desperate with no other options, and they wished it could be different. It was especially courageous of those farmers to share these sentiments.

For me, investigating farmworker conditions has meant coming to understand that the small farmers who employ them were not immune to labor management practices designed to extract the most from their workers for the least amount of remuneration. Because of the intimacy of regional farming, I was able to see how the price of proximity on local farms was borne by workers who had complicated relationships with paternalistic employers. In addition, I learned how Hudson Valley farmers actively recruited immigrant workers whom they understood would have little alternative but to accept the conditions on offer due to their legal status.

I hope this book will be read as a forceful invitation for scholars and others to see the explicit connections between three topics that are usually discussed separately: immigration, labor conditions for all U.S.-based workers, and the food we eat. Although my conclusions are drawn from a regional study of the agricultural industry, they apply not only to farming nationwide but also to the food processing and preparation industries, including meat processing factories, grocery stores, and every class of restaurant. There should be a strong alignment between the fulfillment of consumer expectations regarding sustainable foods and the improvement of labor conditions for U.S. citizens and immigrant workers alike. Historically, one of the main obstacles to the latter is that citizen workers have a hard time seeing how the betterment of their economic futures is tied to doing the same for noncitizen workers. The concentrated power of employer influence on state policy making has also stood in the way.

In recent years the alternative food movement has struck hard at the legitimacy of the industrial agricultural system. Although far from fatal, this blow has revealed the shared assumptions that bind consumers to a conventional food supply. In that moment of opportunity resides the potential for altering the conversation about the political economy of farm labor. Farmworker advocates can now seize that opportunity to increase their ally base and promote workers' rights to a new and rapidly growing segment of the proactive public. This new approach to pursuing justice could be characterized as the inverse of Upton Sinclair's old maxim: farmworker advocates are now aiming at the public's stomach in the hope of hitting its heart.

Methodological Appendix

THIS STUDY IS THE RESULT of inductive research; I did not start out with hypotheses to test, nor were my research goals predetermined, aside from a general desire to document the work lives of this new Latino population of farmworkers as well as advocates' efforts to improve their living and working conditions. My data analysis involved close readings of my interviews and field notes, during which I identified common themes and correspondingly coded my data. This also applied to the organizational material (including pamphlets, press releases, internal evaluations, fundraising letters, newsletters, and reports) that I collected from the nonprofits I studied, including farmer organizations. In addition to approaching farm labor concerns from different standpoints, I also triangulated my research by studying journalistic accounts, government information, nonprofit reports, management texts, and scholarly research on agriculture and farm labor in the Hudson Valley and New York. Case studies often provide deep, detailed studies as opposed to more easily generalizable data; my multimethod approach aimed to mitigate any reliability issues related to the small number of interviews compared with their respective populations. In particular, I limited the number of interviews with farmers because I quickly reached a data saturation point in my conversations with them, and also because of the availability of census, government, nonprofit, and scholarly data, including other surveys and interviews with Hudson Valley and New York farmers that corroborated my results.

My interview data include 113 structured, open-ended interviews with Hudson Valley farmworkers in 2002, along with thirty-five repeat interviews conducted in 2008 and 2009. All but a few interviews took place in labor camps located directly on the nineteen farms where they worked. To locate farmworkers, I identified high-acreage farms in the Hudson Valley that had crops for fall harvest through listings from the New York Department of Agriculture and Markets. Interviewees and others I met on the farms helped me find additional workers to interview—a classic snowball technique. Staff at farmworker advocacy organizations introduced me to workers on four farms. In addition, in 2008 I interviewed a dozen former farmworkers living in or near Hudson and New Paltz. All of these interviews were confidential. In addition to these formal interviews (lasting from forty-five to ninety minutes), I met and spoke with dozens of farmworkers from across the state in the course of my participant observation with the Justice for Farmworkers campaign.

Interviews with others—including twenty-four New York State farmworker advocates whom I met through farmworker-related events and meetings—were semistructured and open-ended and took place between 2003 and 2009. My farmer sample of eighteen Hudson Valley growers was generated by contacting the owners of the farms where I interviewed workers, networking at the Union Square farmers' market in Manhattan, and through referrals from other farmers. They were interviewed in 2003 and 2009. Moreover, I conducted interviews with eight New York State Department of Labor representatives and twenty-one state legislators and aides, lobbyists, representatives from farmer organizations, and farmworker service providers. All interviews were confidential except for three advocates who were executive directors of their organizations.

Participant observation for this study, from 2001 and 2010, involved more than a hundred advocacy-related meetings, farmers' markets, food and farm conferences, and active attendance at dozens of public events—rallies, marches, delegations, public forums, fundraising events, community meetings, and legislative sessions and hearings. Between 2003 and 2009 I attended numerous strategy meetings for the Justice for Farmworkers campaign as a member of the campaign, and I also attended board meetings and strategy sessions of the Rural and Migrant Ministry as a board member. Although I did not draw directly on these meetings due to confidentiality agreements, they provided important context and leads for my research. In addition, I spent a week working with the Centro Independiente de Trabajadores Agrícolas (CITA), also known as the Independent Farmworker Center, in the summer of 2003, and on countless occasions I accompanied the staff of Rural and Migrant Ministry as they worked in their office, visited farms, met with consultants, and implemented a legislative campaign in Albany. In the fall of 2009 I spent a week with Jim Schmidt, the former executive director of Farmworker Legal Services of New York. I also engaged in numerous informal conversations, conference calls, and email exchanges as a member of the Justice for Farmworkers Campaign and board member of the Rural and Migrant Ministry. Some of the testimony I present was collected during the course of my participant observation, when remarks were public.

NOTES

INTRODUCTION

1. Sinclair, "What Life Means to Me."

2. Guthman, *Agrarian Dreams;* Harrison, "Lessons Learned from Pesticide Drift."

3. Brown and Getz, "Privatizing Farm Worker Justice."

4. See, e.g., Walsh, "America's Food Crisis and How to Fix It."

5. Black, "Barry Estabrook's 'Tomatoland,' an Indictment of Modern Agriculture."

6. Sullivan, "Columnist Nicholas Kristof."

7. Between 1989 and 2000 the state saw the employment of black farmworkers decline from half the agricultural work force to one-quarter, while the proportion of Latino farmworkers increased from one-third to two-thirds. Pfeffer and Parra, *Immigrants and the Community.*

8. State laws follow the precedent of the National Labor Relations Act of 1935 (NLRA), also known as the Wagner Act, which was created to address unfair labor practices. This federal law gives most private sector workers these rights and obliges employers to recognize and bargain with certified unions. Farmworkers, domestic workers, and others, however, are excluded from the protections of the act. For more information, see Gold, *An Introduction to Labor Law, Revised Edition;* Fojo, Burtness, and Chang, *Inventory of Farmworker Issues and Protections in the United States.*

9. Field sanitation (toilet and hand-washing facilities) "shall be located within a one-quarter mile walk of most hand-laborers or at the closest point that may be accessible by motor vehicle." New York State Consolidated Laws, Chapter 31 Labor Law, Article 7, § 212-D.

10. In *The Power to Choose,* Naila Kabeer similarly offers multiple accounts of Bangladeshi women's working lives, including those of workers themselves, advocates, journalists, employers, and scholars.

11. New York farmworkers were the focus of a handful of reports between the 1940s and 1990s. See, in particular, Amidon, *What's Next for New York's Joads?;*

Barr, *Liberalism to the Test*; Edid, *Agricultural Labor Markets in New York State and Implications for Labor Policy*; Mendez and Diaz, *Separate & Unequal*. There was little to no literature on the Latino population of New York farmworkers at the time I began my own research in 2000. Since 2004, New York's newer Latino population of farmworkers has received some scholarly attention. See Gray and Hertel, "Immigrant Farmworker Advocacy"; Gray, "How Latin American Inequality Becomes Latino Inequality"; Gray, "Mechanics of Empowerment"; Gray, *The Hudson Valley Farmworker Report*; Maloney and Grusenmeyer, *Survey of Hispanic Dairy Workers in New York State*; Maloney, "Understanding the Dimensions of Culture"; Pfeffer and Parra, "Strong Ties, Weak Ties and Human Capital"; Pfeffer and Parra, *Immigrants and the Community*; Pfeffer and Parra, *Immigrants and the Community: Community Perspectives*; Pfeffer and Parra, *Immigrants and the Community: Farmworkers with Families*; Pfeffer and Parra, *Immigrants and the Community: Former Farmworkers*; Pfeffer and Parra, "New Immigrants in Rural Communities." So too, while research on farmworker advocacy in the United States has been the subject of many scholarly works, there is only one study that focuses on northeastern farmworkers, a book about Puerto Rican farmhands in New Jersey. See Bonilla-Santiago, *A Case Study of Puerto Rican Migrant Farmworkers Organizational Effectiveness in New Jersey*. This useful but decades-old study is a social movement analysis of a grassroots organization and the means by which it was able to improve conditions for New Jersey farmworkers.

12. Diamond and Soto, *Facts on Direct-to-Consumer Food Marketing*.

13. United States Department of Agriculture, National Agricultural Statistics Service, *Table 1. Historical Highlights*.

14. In this way my work answers the call for a regionally specific analysis of labor markets (Peck, *Work Place*) and agroecological problems (Niles and Roff, "Shifting Agrifood Systems").

15. Indeed, the presence of new, mostly male, agricultural workers is usually "ephemeral, limited, and frequently almost invisible." Zúñiga and Hernández-León, "Introduction," xiii. As such, they are unlikely to attract the same attention as their urban and suburban counterparts, who are concentrated in more densely populated neighborhoods.

16. Marion Nestle and W. Alex McIntosh reflect on food and food movement books to develop a core reading list in "Writing the Food Studies Movement." In addition, Daniel Niles and Robin Jane Roff offer a review of the literature that they term "agrifood studies" in "Shifting Agrifood Systems." Food writing in this area can broadly be categorized as activist/popular or academic, with the understanding that there is significant overlap. Much of this writing is focused on the implications of our food choices. Important works in the activist/popular category relevant to this study are Belasco, *Appetite for Change*; Gussow, *This Organic Life*; Halweil, *Eat Here*; Kingsolver, *Animal, Vegetable, Miracle*; Klinkenborg, *Making Hay*; Lappé and Terry, *Grub*; Lappé and Lappé, *Hope's Edge*; Lappé, *Diet for a Small Planet*; McKibben, *Deep Economy*; Nestle, *What to Eat*; Patel, *Stuffed and Starved*; Petrini, *Slow Food (The Case for Taste)*; Pollan, *The Botany of Desire*; Pollan, *The Omnivore's*

Dilemma; Pringle, *Food Inc.;* Rifkin, *Beyond Beef;* and Schlosser, *Fast Food Nation.* In the academic category, prominent works include Allen, *The Human Face of Sustainable Agriculture;* Allen, *Together at the Table;* Guthman, *Agrarian Dreams;* Kimbrell, *The Fatal Harvest Reader;* Lyson, *Civic Agriculture;* Magdoff, Foster, and Buttel, *Hungry for Profit;* Nabhan, *Coming Home to Eat;* Nestle, *Food Politics;* Shiva, *Stolen Harvest;* Wirzba, *The Essential Agrarian Reader.* Other works bridge the activist/popular and academic categories, such as Alteri, *Agroecology* and Gussow, *The Feeding Web.* In addition, the writings of Wendell Berry *(The Unsettling of America* and *What Are People For?)* and Aldo Leopold *(Sand County Almanac)* have greatly shaped the alternative food movement in terms of land use and conservation.

17. For scholars that call for an inclusion of farmworkers in food systems research, see Allen, "Mining for Justice in the Food System"; Brown and Getz, "Toward Domestic Fair Trade?"; Garcia, "Labor, Migration, and Social Justice in the Age of the Grape Boycott"; Guthman, *Agrarian Dreams;* Allen, *Together at the Table.* In the popular literature, Eric Schlosser's consistent treatment of farm labor is a rare exception. See his *Fast Food Nation, Reefer Madness,* and "Penny Foolish." In 2006 the first special food issue of the *Nation,* the leading U.S. left-wing magazine, addressed labor issues, but the follow-up food issue in 2009 neglected the topic. How the expanding organic sector intersects with farm labor, particularly in California, has been the focus of some studies, including Guthman, *Agrarian Dreams;* and Schreck, Getz, and Feenstra, "Farmworkers in Organic Agriculture."

18. Nestle and McIntosh, "Writing the Food Studies Movement," 164.

19. Holbrook and Corfman, "Quality and Value in the Consumption Experience."

20. In food industry studies, Steve Striffler's study pushes for integrating labor issues with consumer power, examines the exploitation and stories of new immigrants, and investigates shifts in the poultry industry's labor market, yet he proposes that smaller, regional producers offer better work conditions without a thorough examination of that sector. See Striffler, *Chicken.*

21. Moss and Tilly, *Stories Employers Tell;* Steinberg, "Immigration, African Americans, and Race Discourse"; Waldinger and Lichter, *How the Other Half Works.*

22. Guthman, *Agrarian Dreams,* 12.

I. AGRARIANISM AND HUDSON VALLEY AGRICULTURE

1. Novesky and Crawshaw, "Branding the Region." Minetta Brook, an arts organization, and the Glynwood Institute for Sustainable Food and Farming held various sessions with diverse participants in 2000 to "explore the existing potential connection between food, cuisine and the Valley's identity." Two of those sessions were "Could the Hudson Valley be the Next Napa?" and "Strengthening the Regional Food System in the Hudson Valley."

2. Kingsolver, *Animal, Vegetable, Miracle*, 17, 20. Some select and disparate examples of food culture might include China's four regional culinary systems, the Provençal and Punjabi food traditions, the Italian "Slow Food" movement, and the Bay Area's viticulture-inspired gourmet and locally based cuisine. Food culture may also be defined by the particular social role of food or can be more specifically related to prize ingredients, cooking techniques, and special occasion meals. It is also related to the French concept of *terroir*, roughly translated as "a sense of place," which refers to how a particular regional ecology (natural, historical, cultural, etc.) influences the production and consumption of food and drink. For an examination of *terroir* and an analysis of the particular regional food ecologies of not only France but also California, Wisconsin, and Vermont, see Trubek, *The Taste of Place*.

3. Kummer, "The Great Grocery Smackdown."

4. DiNapoli and Bleiwas, *Farmers' Markets in New York City*.

5. www.ediblecommunities.com.

6. Gassan, *The Birth of American Tourism*.

7. Ibid.

8. Toole, "The Role of Agriculture at Hudson Valley Historical Sites."

9. Gary Nabhan *(Coming Home to Eat)* defines "local" as being within a 250-mile radius of his home; Eatlocalchallenge.com and Localdiet.org have proposed a hundred miles; and Joan Gussow *(This Organic Life)* proposes "within a day's leisurely drive of our homes." According to a USDA report, "'local food' is an ambiguous characteristic." Instead of relying on how farm food travels from farm to table, the authors propose defining "local" by how the food is marketed. Thus, both direct-to-consumer and intermediated (direct-to-grocer/restaurant) food sales are considered "local foods" in their study. Low and Vogel, *Direct and Intermediate Marketing of Local Foods in the United States,* 1. Low and Vogel incorporate Hand and Martinez, "Just What Does Local Mean?" and Martinez et al., *Local Food Systems.*

10. www.wholefoodsmarket.com/locally-grown.

11. Jennifer Maiser, "10 Reasons to Eat Local Food," April 8, 2006 at www.life-beginsat30.com/elc/2006/04/10_reasons_to_e.html.

12. Born and Purcell, "Avoiding the Local Trap."

13. Feenstra, "Local Food Systems and Sustainable Communities"; Norberg-Hodge, Merrifield, and Gorelick, *Bringing the Food Economy Home;* Pacione, "Local Exchange Trading Systems."

14. Murdoch, Marsden, and Banks, "Quality, Nature, and Embeddedness."

15. Lyson, *Civic Agriculture*. Other scholars critique the a priori assumptions about the benefits of alternative agriculture systems. See Allen et al., "Shifting Plates in the Agrifood Landscape"; DuPuis and Goodman, "Should We Go 'Home' to Eat?"; Hinrichs et al., *Moving Beyond Global and Local;* Hinrichs, "The Practice and Politics of Food System Localization"; Weatherell, Tregear, and Allinson, "In Search of the Concerned Consumer"; Winter, "Embeddedness, the New Food Economy and Defensive Localism." For a more substantial review of the assumptions about local food systems, see Born and Purcell, "Avoiding the Local Trap," who review the

literature in terms of ecological sustainability, social and economic justice, and food quality and human health. It is important to note that the food studies dialogue about the meanings of scale and the distribution of local food is strongly influenced by geographers, in particular David Harvey *(Justice, Nature and the Geography of Difference)*.

16. Schumacher, *Small Is Beautiful*.

17. Flinn and Johnson, "A Comparison of Agrarianism in Washington, Idaho, and Wisconsin"; Buttel and Flinn, "Sources and Consequences of Agrarian Values in American Society"; Carlson and McLeod, "A Comparison of Agrarian Values in American Society"; Singer and de Sousa, "The Sociopolitical Consequences of Agrarianism Reconsidered"; Molnar and Wu, "Agrarianism, Family Farming, and Support for State Intervention in Agriculture."

18. Johnstone, "Old Ideals versus New Ideas in Farm Life"; Rohrer and Douglas, *The Agrarian Transition in America*.

19. Jefferson, *Notes on the State of Virginia*, 176.

20. Crèvecoeur, *Letters from an American Farmer*. Although agrarianism was fuel for populism, its roots are literary, and it draws on classical writers. Thompson, in "Agrarianism and the American Philosophical Tradition," calls Thomas Jefferson the "patron saint of U.S. agrarianism" (3). Agrarianism draws on the writings of Aristotle, Hesiod, Xenophon, Cato, Cicero, Virgil, Horace, Locke, Montesquieu, and more. See Griswold, *Farming and Democracy;* Brewster, *A Philosopher among Economists;* Hofstadter, *The Age of Reform*. In American writing, along with Jefferson and Crèvecoeur, Thomas Paine, Philip Freneau, Hugh Henry Brackenridge, and George Logan influenced the tradition. On the prevalence of agrarianism in the eighteenth-century United States, see Eisinger, "The Freehold Concept in Eighteenth-Century American Letters."

21. This paragraph draws on the work of Hofstadter, *The Age of Reform*, and Johnstone, "Old Ideals versus New Ideas in Farm Life." It is worth noting that there is no definitive version of agrarianism. Rather, its definition has been revised over time by those who have employed the concept. See Flinn and Johnson, "Agrarianism among Wisconsin Farmers."

22. Johnstone, "Old Ideals versus New Ideas in Farm Life," 132.

23. Hofstadter, *The Age of Reform*, 29.

24. Johnstone, "Old Ideals versus New Ideas in Farm Life"; Hofstadter, *The Age of Reform*.

25. Hofstadter, *The Age of Reform*, 28.

26. Ibid., 38.

27. Rohrer, "Agrarianism and the Social Organization of U.S. Agriculture."

28. Hofstadter, *The Age of Reform;* Paarlberg, *Farm and Food Policy;* Rohrer and Douglas, *The Agrarian Transition in America*. Romantic thinkers like Emerson and Thoreau, though they followed a different track, further elevated the social creed of rural production.

29. Ways, *The Negro and the City*, 22.

30. Dolan, "New Detroit Farm Plan Taking Root."

31. Beus and Dunlap, "Endorsement of Agrarian Ideology and Adherence to Agricultural Paradigms," 481.

32. See, for example, Thompson, "Agrarianism and the American Philosophical Tradition"; Buttel and Flinn, "Sources and Consequences of Agrarian Values in American Society"; McConnell, *The Decline of Agrarian Democracy;* Beus and Dunlap, "Endorsement of Agrarian Ideology and Adherence to Agricultural Paradigms"; Berry, *The Unsettling of America;* Merrill, *Radical Agriculture;* Buttel, "Agriculture, Environment, and Social Change"; Burch Jr., *Daydreams and Nightmares;* Heffernan and Heffernan, "Impact of the Farm Crisis on Rural Families and Communities"; Johnstone, "Old Ideals versus New Ideas in Farm Life"; Hofstadter, *The Age of Reform.*

33. Pedersen argues that four historical factors account for U.S. agriculture's influence on the law: 1) the extensive use and role of the land in agriculture, 2) the seasonal nature of agriculture, 3) atomistic production (many producers without vertical integration), and 4) agricultural exceptionalism. Pedersen, "Introduction to the Agricultural Law Symposium," 409.

34. Danbom, "Romantic Agrarianism in Twentieth-Century America," is referring to Lewis, *The American Adam.*

35. Paarlberg, *Farm and Food Policy.*

36. Berry, "The Whole Horse," 238.

37. Carlson, "Agrarianism Reborn"; Sheingate, "Institutions and Interest Group Power." For example, in a 2004 op-ed in the *New York Times,* Victor Davis Hanson, a writer, farmer, and senior fellow at the Hoover Institution, wrote, "Agriculture is more than just feeding people; it is the historic center of bedrock American social and cultural values.... In this most dangerous period in our nation's history, agriculture remains our most precious resource." Hanson, "A Secretary for Farmland Security." See also Hanson, *Fields without Dreams,* and Hanson, *The Other Greeks,* which espouses conservative, nativist, and libertarian ideas alongside agrarian romanticism and laments the loss of a rooted culture that is displaced by the consolidation of farming. Agrarianism, however, is not just the domain of conservatives; *New York Times* columnist Nicholas Kristof, in an article criticizing corporate monoculture farming and the government's support for it, tapped into his own childhood experiences on a farm to argue that "the family farm traditionally was the most soulful place imaginable." Kristof, "Food for the Soul."

38. Hofstadter, *The Age of Reform;* Beus and Dunlap, "Endorsement of Agrarian Ideology and Adherence to Agricultural Paradigms"; Johnstone, "Old Ideals versus New Ideas in Farm Life."

39. Martin, "Mexican Workers and U.S. Agriculture."

40. United States Department of Labor, *Migrant Farmworkers,* 40.

41. Guthman, *Agrarian Dreams,* 177.

42. Food Security Act of 1985, Public Law 99–198, HR 2100, 99th Cong., 1st sess., Congressional Record 131 (October 8), H 8461.

43. Hoppe et al., "Structure and Finances of U.S. Farms," 2.

44. United States Department of Agriculture, National Agricultural Statistics Service, *Table 1*, 72.

45. Dun & Bradstreet, *Comprehensive Business Report.*

46. United States Department of Agriculture, National Agricultural Statistics Service, *Table 1*, 72–73. Judging a farm's labor needs by acreage can be misleading since orchards are very concentrated. For example, a farmer with a two-hundred-acre orchard explained to me, "This is a fairly good size for apples, small if it was row crops. It takes a lot of people for two hundred acres of apples, whereas for two hundred acres of farm crops, you just need a few guys."

47. Conford, *The Origins of the Organic Movement;* Guthman, *Agrarian Dreams.*

48. Guthman, *Agrarian Dreams,* 12.

49. Ibid., 57.

50. Ibid., 174.

51. Mitchell, *The Lie of the Land;* Lewthwaite, "Race, Paternalism, and 'California Pastoral.'"

52. Lewthwaite, "Race, Paternalism, and 'California Pastoral,'" 9.

53. New York Farm Bureau, *Letter Regarding Farmworkers March from Auburn to Albany.*

54. Lyson and Green, "The Agricultural Marketscape."

55. Kingsolver, *Animal, Vegetable, Miracle,* 123.

56. Some scholars have combed through census or archival data to re-create the lives of common family farmers. See Bruegel, *Farm, Shop, Landing;* Humphrey, *Land and Liberty Hudson Valley Riots in the Age of Revolution;* Horne, "Life on a Rocky Farm"; and the Putnam County Historical Society's survey and exhibit on the Oakley Farm. But these accounts largely neglect hired hands. Studies that do include workers, re-creating history based on census, farm diaries, and other archival data, include McDermott, *Dutchess County's Plain Folks;* and Parkerson, *The Agricultural Transition in New York State.*

57. Amidon, *What's Next for New York's Joads?;* Barr, *Liberalism to the Test;* Close, *The Joads of New York;* Friedland, "Labor Waste in New York"; Friedland and Nelkin, *Migrant;* García-Colón, "Claiming Equality"; Nelkin, *On the Season.*

58. Maloney and Grusenmeyer, *Survey of Hispanic Dairy Workers in New York State;* Maloney, "Understanding the Dimensions of Culture"; Margolies, *Training Needs Assessment of Farm Workers in Orange and Sullivan Counties, NY;* Pfeffer and Parra, *Immigrants and the Community;* Pfeffer and Parra, *Immigrants and the Community: Community Perspectives;* Pfeffer and Parra, *Immigrants and the Community: Farmworkers with Families;* Pfeffer and Parra, *Immigrants and the Community: Former Farmworkers;* Pfeffer and Parra, "New Immigrants in Rural Communities"; Pfeffer and Parra, "Strong Ties, Weak Ties and Human Capital."

59. Humphrey, *Land and Liberty.*

60. Bruegel, *Farm, Shop, Landing,* 20.

61. Farmers detailed such labor exchanges in their diaries. See, for example, Parkerson, *The Agricultural Transition in New York State.*

62. Bruegel, *Farm, Shop, Landing*, 21, 110; McDermott, *Dutchess County's Plain Folks*.

63. Bruegel, *Farm, Shop, Landing*. Slaves were vital to the early economy of New York agriculture, and in the mid-eighteenth century the state had more slaves than any other nonplantation colony. Nonetheless, New York agriculture in the postcolonial era was not structured around slave labor. The first census conducted in 1790 recorded 21,324 slaves in the state, about 6 percent of the general population. These numbers would decline substantially as New York encouraged the freeing of slaves through 1799 legislation and required it by 1827. One exception was that the children of slaves had to reach a certain age before manumission and thus might not be freed until after 1827. The 1799 Gradual Manumission Act freed children born to slave women when males reached twenty-eight and women reached twenty-five. Laws of New York, 1799, LXII, and 1817, CXXXVI.

64. Bruegel, *Farm, Shop, Landing*; Danielsson, *To Celebrate the Land*; McDermott, *Dutchess County's Plain Folks*.

65. Bruegel, *Farm, Shop, Landing*, 217.

66. Ibid., 10.

67. Ellis, "Land Tenure and Tenancy in the Hudson Valley, 1790–1860," 75.

68. McCurdy, *The Anti-Rent Era in New York Law and Politics, 1839–1865*, xiii.

69. McDermott, *Dutchess County's Plain Folks*, 172.

70. Bruegel, *Farm, Shop, Landing*, 144–45.

71. Johnstone, "Old Ideals versus New Ideas in Farm Life," 149.

72. McDermott, *Dutchess County's Plain Folks*, 153.

73. Bruegel, *Farm, Shop, Landing*, 220.

74. Ibid., 250 n. 34.

75. One such farmhand, George Atkins, profiled by historian William McDermott, lived with his family and worked in Dutchess County in the mid-nineteenth century (McDermott tracks him from 1840 to 1875 through census data and a storekeeper's records). McDermott's reconstruction of the Atkins family's life is a welcome exception to the more typical neglect of laborers' lives in the chronicles of Hudson Valley history. Atkins's employment trajectory was from farm laborer, day laborer, and sometimes mill worker, with additional employment by the local general store owner, to becoming a miller for three years (which offered highly desirable year-round work). At age sixty-five he was once again a farm laborer when he became too old to work at the mill but still had the energy to tend to crops. Atkins arrived with his family in Dutchess County with few personal possessions, perhaps only beds, chairs, clothes, and few kitchen items. He probably never ventured more than three miles from his home, and although he may have raised pigs and produced spirits, his family's quotidian experiences are best described as hand to mouth. Their few splurges included tobacco, a onetime firecracker purchase, and three cents of candy on Christmas Eve 1850, in all likelihood for Atkins's eleven-year-old son, Billy. McDermott details the meager existence of farm and general laborers and argues that the lives of such workers were marked by uncertainty and insecurity. McDermott, *Dutchess County's Plain Folks*.

76. Ibid., 3–4.

77. Zimmerman, "Nineteenth Century Wheat Production in Four New York State Regions."

78. Hahamovitch, *The Fruits of Their Labor*, 26.

79. Bayne, *County at Large*.

80. Horne, "Life on a Rocky Farm," 19.

81. Johnstone, "Old Ideals versus New Ideas in Farm Life," 147.

82. Hofstadter, *The Age of Reform*, 120.

83. Johnstone, "Old Ideals versus New Ideas in Farm Life."

84. Danielsson, *To Celebrate the Land*.

85. For details on the Italian padrone system, see Hahamovitch, *The Fruits of Their Labor*, chapter 2.

86. Hurd, *New York's Harvest Labor*, 4.

87. Stanton, *The Changing Landscape of New York Agriculture in the Twentieth Century*, 12–13.

88. Hahamovitch, *The Fruits of Their Labor*, 7.

89. Ibid., 103–12. One group the state would not organize for farm work—at least not until the end of the war, and only then as a contingency work force—was women, as many Americans considered it inappropriate for white women who were not part of the working class to toil in the fields. Despite ensuing controversy, female college students and women from the "leisure class" were volunteer laborers who were organized and trained by women's groups, colleges, and the Women's Land Army—all private initiatives. For more general information on the Women's Land Army, see Carpenter, *On the Farm Front*. Women again engaged in wartime farm work during World War II. One of the farmers I interviewed recalled teams of "farmerettes" being sent up from Hunter and Fordham colleges. In addition, local youth aged eleven to seventeen, including those trained through the Farm Cadet Victory Corps, were released from school obligations to work the harvest. Brooklyn, Hunter, City, and Queens colleges all sent students to farms in Dutchess County in 1942. In 1943, Brooklyn College placed students in the Farm Cadet Victory Corps for work on farms in upstate New York. An oral history project of the latter details students' experiences in Morrisville in Madison County. See Back, *Student Voices from World War II and the McCarthy Era*.

90. Hahamovitch, *The Fruits of Their Labor*.

91. Stanton, *The Changing Landscape of New York Agriculture in the Twentieth Century*, 12.

92. Hurd, *New York's Harvest Labor*. This early-twentieth-century migration would create a labor scarcity in the South that accounted for better treatment of black workers. Williams, *The Negro Exodus from the South*, quoted in Foner and Lewis, eds., *Black Workers*, 307. See also Hahamovitch, *The Fruits of Their Labor*.

93. Stanton, *The Changing Landscape of New York Agriculture in the Twentieth Century*, 13.

94. Ibid., 12.

95. Except where noted, the data for this paragraph is from Hurd, *New York's Harvest Labor.*

96. A document describing archival data on the New York State War Council's Farm Manpower Service lists correspondence regarding "the possibility of recruiting laborers from non-traditional sources, such as state mental health facilities." Norris and Engst, *Migrant Farmworkers Records in Upstate New York.* Another report mentions "inmates of institutions." Hurd, *New York's Harvest Labor,* 3. It is worth noting that the report uses the term "volunteer seasonal workers" to describe all laborers, including prisoners of war.

97. Such a practice still occurs today because foodies are curious about how their food is grown. A Long Island organic farmer told me he often hosted city vacationers who worked on his farm.

98. Also in the New York State War Council archive, a summary description of photographs includes, "Chinese labor as part of the Farm Manpower Service." Norris and Engst, *Migrant Farmworkers Records in Upstate New York,* 53. These workers were probably recruited from New York City.

99. For an examination of German prisoner of war camps in New York during World War II, see Mazuzan and Walker, "Restricted Areas."

100. For a more comprehensive discussion of the government's role in supplying labor to farms, see Hahamovitch, *The Fruits of Their Labor,* especially chapter 7.

101. Hurd, *New York's Harvest Labor.*

102. Amidon, *What's Next for New York's Joads?,* 8.

103. Close, *The Joads of New York.*

104. "Interstate" migrants numbered 2,296 in 1948; "Southern migrants" numbered 16,661. Hurd, *New York's Harvest Labor.*

105. Another report by the Consumers League of New York stated that most blacks were from the "Deep South," particularly Florida. Amidon, *What's Next for New York's Joads?*

106. For an examination of racism against black farmworkers in the pre–World War II era, see Hahamovitch, *The Fruits of Their Labor.* For a description of ethnic and racial characteristics, talents, and problems of workers, see Adams, *Farm Management,* chapter 22.

107. Stanton, *The Changing Landscape of New York Agriculture in the Twentieth Century,* 12.

108. Griffith and Kissam, *Working Poor;* Thomas-Lycklama à Nijeholt, *On the Road for Work.*

109. This number is from the New York State Employment Service. See Nelkin, *On the Season,* 3. A 1959 demographic study of New York's black migrants showed that 18 percent were under the age of fourteen, and 15 percent were between the ages of fifteen and nineteen. Larson, *Migratory Agricultural Workers in the Eastern Seaboard States, Rural Poverty in the United States,* quoted in Nelkin, *On the Season,* 4.

110. Puerto Ricans had been brought to the United States for agricultural work beginning in the early 1900s, but it was not until the 1940s that their recruitment became more official through a joint program of the Puerto Rican and U.S. Depart-

ments of Labor. In 1948 New York hired 1,051 Puerto Ricans from Puerto Rico (as opposed to New York City, where Puerto Ricans were also recruited). See Hurd, *New York's Harvest Labor*, 6. By 1953 New York employed 3,000 Puerto Ricans in agriculture. See Mirengoff, *Puerto Rican Farm Workers in the Middle Atlantic States*, 1. In the surrounding states, including Pennsylvania, New Jersey, and Massachusetts, Puerto Ricans developed a more concentrated ethnic niche in agriculture. For research on Puerto Rican farmworker collective action in New Jersey, see Bonilla-Santiago, *A Case Study of Puerto Rican Migrant Farmworkers Organizational Effectiveness in New Jersey*. And, for an example of Puerto Rican farmworker protest in upstate New York, see García-Colón, "Claiming Equality."

111. Barr, *Liberalism to the Test*, 5; Nelkin, *On the Season*, 3.

112. U.S. President's Commission on Migratory Labor, *Migratory Labor in American Agriculture*.

113. Stanton, *The Changing Landscape of New York Agriculture in the Twentieth Century*, 13.

114. Ibid., 7–9.

115. Daly, *Agriculture in Transition*, 1.

116. Ibid., 2–3.

117. The change happened over time. In 1987 there were 105 dairy farms in Dutchess County, by 1997 51, by 2007 39, and in 2009 about 20 to 25. Data from United States Department of Agriculture, National Agricultural Statistics Service, *Table 1*, and personal correspondence with Les Hulcoop of Cornell Cooperative Extension, October 27, 2009.

118. United States Department of Agriculture, *2007 Census of Agriculture*.

119. Sorenson, Green, and Russ, *Farming on the Edge*.

120. Though Suffolk County in New York enacted the first program for the Purchase of Development Rights (PDR) as early as 1974, it was not until the 1990s that governmental and nonprofit initiatives were enacted to protect open space and working farms from housing developments and suburban sprawl.

121. Ferguson, *Saving Working Landscapes*, 5.

122. Such an analysis would need to include all affiliated employment such as transportation and veterinarians.

123. United States Department of Agriculture, National Agricultural Statistics Service, *Table 1*.

124. New York is ranked second nationally in the production of apples, pumpkins, and maple syrup; third for wine and juice grapes, cauliflower, cabbage, and corn silage; fourth for tart cherries, pears, fresh market sweet corn, squash, dairy, and fresh market snap beans; and fifth for onions and fresh market vegetables. New York also ranks as seventh for cucumbers and floriculture, eighth for strawberries, and eleventh for tomatoes (www.agriculture.ny.gov/agfacts.html).

125. "Wal-Mart Could Get Wounded in Grocery Wars."

126. Oxfam America, *Like Machines in the Fields*; Jayaraman, *The Hands That Feed Us*.

127. Toole, "The Role of Agriculture at Hudson Valley Historical Sites."

2. THE WORKERS

1. Integrated Pest Management (IPM) is a popular system dedicated to using only the minimum volume of chemicals necessary. One farmer I interviewed posted information about IPM on his farm website due to the number of customers inquiring about his pesticide practices.

2. Pollan, *The Omnivore's Dilemma*, 8.

3. Kingsolver, *Animal, Vegetable, Miracle*, 306.

4. See Schlosser's *Reefer Madness* and *Fast Food Nation*.

5. United States Department of Agriculture, *2007 Census of Agriculture*.

6. See the Methodological Appendix for details on the farms where I conducted interviews.

7. I include a few workers involved in food preparation because the workers engaged in this task were employed by a grower and lived on a farm. Sixty-four percent of farmworkers undertook harvesting, whether picking fruit or cutting vegetables, and almost one-quarter of these respondents also worked during the planting season. On two different farms, three workers reported working at farmers' markets in addition to their farm tasks.

8. Pfeffer and Parra, "New Immigrants in Rural Communities."

9. Workers reported average annual incomes from only farm work in the Hudson Valley of $6,643 in 2002 and $7,345 in 2001.

10. These are rough estimates reported by the workers. The discrepancy between 2001 and 2002 is because of a significant crop loss due to weather damage in 2002.

11. In 2002 the U.S. Department of Health and Human Services poverty guidelines were $8,860 for one person, $11,940 for two people, and $15,020 for three people. United States Department of Health and Human Services, "The 2002 HHS Poverty Guidelines."

12. Jayaraman, *The Hands That Feed Us*, 37.

13. Those making less than $6 an hour accounted for 11 percent of the sample. Twenty-six percent of interviewees earned between $7 and $8 an hour, and 14 percent earned $8 or more.

14. Twenty-two percent of workers regularly worked more than six days a week, 14 percent worked seven days a week, and 8.3 percent worked six days a week plus part of the seventh day. An additional 7.4 percent reported that in addition to working six days a week, they sometimes worked the seventh day.

15. Jayaraman, *The Hands That Feed Us*, 63.

16. For a more detailed examination of Hudson Valley farmworkers' living and working conditions, as well as their opinions on current policy debates, see Gray, *The Hudson Valley Farmworker Report*.

17. New York's 1937 Labor Relations Act established that farmworkers were not "employees." Subsequently, the 1938 New York State Constitution excluded them from rights enjoyed by other workers in the state, including the right to overtime pay, a day of rest, and collective bargaining protections. Although these rights are elusive for farmworkers in most other states, Hawaii was an early exception, granting

collective bargaining protections in the state's 1945 Employment Relations Act, while California amended its labor law in 1975 to require overtime pay and collective bargaining protections. Today eleven states provide collective bargaining protections and six require overtime pay.

18. On a federal level, for those under the age of twenty there is a youth minimum wage of $4.25 that lasts for the first ninety consecutive days of employment in any job. New York State does not recognize the youth minimum wage provisions, and youth workers in most jobs there earn the same minimum wage as adults. Personal correspondence with Irv Miljoner of the United States Department of Labor, Long Island District Office, June 11, 2013. In contrast, the minimum wage for farmworker youth under the age of sixteen is $3.20 an hour for harvest or nonharvest work. For sixteen- and seventeen-year-old harvest workers, the minimum wage is $3.60 for their first season and $3.80 for their second season. For sixteen- and seventeen-year-old nonharvest workers, the minimum wage is $3.60 an hour for the first three hundred hours and $3.80 per hour for the second three hundred hours. Some tasks, such as operating a feed grinder or forklift and handling certain chemicals, are considered too dangerous for child farmworkers and are prohibited by the state. For child farmworkers there are fewer limits on work hours than in other industries. Twelve- and thirteen-year-old farmworkers are limited to four hours a day. For those fourteen and older there are *no limits on working hours*. For most other youth workers, the limits on work hours are different when school is in session and when school it is not in session. When school is in session, the limit on working hours for fourteen- and fifteen-year-olds in most jobs is three hours on school days and eight hours on other days, with a maximum of eighteen hours a week. Sixteen- and seventeen-year-olds in most jobs have a limit of four hours a day on school days (except Fridays) and eight hours a day on other days, with a maximum of twenty-eight hours a week. When school is in session these youth are permitted to work eight hours a day, with the former having a maximum of forty hours a week and the latter a maximum of forty-eight hours a week. For more details, see "Permitted Working Hours for Minors" at www.labor.ny.gov/formsdocs/wp/LS171.pdf.

19. The influence of growers on the NLRA was not restricted to those of southern provenance, nor were farmworkers excluded only on racist grounds. Farmers nationwide had an all too compelling memory of I.W.W. efforts to undermine growers' power during crucial harvest times in the West and Midwest, and they steadfastly opposed farmworker collective bargaining protections. Linder, *Migrant Workers and Minimum Wages*. For an inside perspective on the passage of the NLRA and the exclusion of farmworkers, see Daniel, *Bitter Harvest*.

20. Edid, *Farm Labor Organizing;* Linder, *Migrant Workers and Minimum Wages;* Rothenberg, *With These Hands*.

21. Earlier legislation had established this pattern at the core of New Deal policies. See Linder, *Migrant Workers and Minimum Wages,* chapter 4; Hahamovitch, *The Fruits of Their Labor,* chapter 6. The 1933 National Industrial Recovery Act (NIRA), designed to give a boost to the Depression-era economy and promote domestic consumption, set minimum wages and maximum hours for all U.S.

workers with the exception of agricultural and domestic workers at the request of southern legislators. Furthermore, the Agricultural Adjustment Act of 1933 allowed for the casual eviction of tenant farmers and sharecroppers.

22. Herbert, "State of Shame."

23. Comment by John Kruspe from Toronto, Canada posted June 9, 2009. Hudson Valley foie gras workers, however, have spoken publicly about their plight, including when they were honored at the 2009 annual Sowing Seeds for Justice Dinner sponsored by the Greater New York Labor-Religion Coalition and Rural and Migrant Ministry.

24. Lichter, "Immigration and the New Racial Diversity in Rural America," 9.

25. Waters and Jiménez, "Assessing Immigrant Assimilation."

26. In New York State, conflict has escalated around day laborer recruitment zones. See *Farmingville,* a documentary film by Carlos Sandoval and Catherine Tambini (P.O.V. 2004), about the response of residents in a small Long Island town to an influx of Mexican day laborers. See also Claffey, "Anti-immigrant Violence in Suburbia"; Semple, "A Killing in a Town Where Latinos Sense Hate."

27. Griffith, *American Guestworkers.* The guest worker programs were created based on growers' claims of labor shortages, yet they were used to disable farmworker collective action. Hahamovitch, *The Fruits of Their Labor;* Majka and Majka, *Farm Workers, Agribusiness, and the State.* Readers may be aware of the West Coast and Southwest Bracero programs (1942–64). See Calavita, *Inside the State;* Galarza, *Merchants of Labor.* The East Coast had complementary programs set up with the Bahamas and Jamaica in 1943 and Puerto Rico in 1944 (counterparts to the West Coast Bracero Program). Although intended to be a temporary wartime program, the Bracero Program continued until 1964, while the guest worker programs from the Caribbean were never terminated and continue offering H-2 agricultural visas. In the 1990s, the H-2 program expanded to include Latin American (predominantly Mexican) workers and also expanded into nonagricultural sectors such as food processing and tourism through contracts spread across the United States. Griffith, *American Guestworkers,* 36. Guest workers are also from Central America, Thailand, Indonesia, and South Africa.

28. Department of Labor Employment and Training Administration, *Labor Certification Process for the Temporary Employment of Aliens in Agriculture in the United States.*

29. Geffert, "H-2A Guestworker Program."

30. See also Griffith, *American Guestworkers;* Oxfam America, *Like Machines in the Fields.*

31. For a brief historical review of union organizing in California from 1900 to 1960, see Ganz, *Why David Sometimes Wins,* chapter 2. For examples of illegal activities to quash farmworker organizing, see Hahamovitch, *The Fruits of Their Labor;* Majka and Majka, *Farm Workers, Agribusiness, and the State;* McWilliams, *Factories in the Field;* Meister and Loftis, *A Long Time Coming.*

32. Majka and Majka, *Farm Workers, Agribusiness, and the State.*

33. Norris, "Industrial Paternalist Capitalism and Local Labour Markets"; Jackman, *The Velvet Glove.*

34. Bennett, "Paternalism"; Alston and Ferrie, *Southern Paternalism and the American Welfare State.*

35. Padavic and Earnest, "Paternalism as a Component of Managerial Strategy."

36. Goodell et al., "Paternalism, Patronage, and Potlatch," 354.

37. Padavic and Earnest, "Paternalism as a Component of Managerial Strategy."

38. Alston and Ferrie, *Southern Paternalism and the American Welfare State.* Paternalism in employer-controlled southern mill towns is examined in Leiter, Schulman, and Zingraft, eds., *Hanging by a Thread.* For a discussion of planter paternalism, see Genovese, *Roll, Jordan, Roll.* For an investigation of how antebellum paternalism shaped planter paternalism, see Alston and Ferrie, *Southern Paternalism and the American Welfare State.*

39. Wingerd, "Rethinking Paternalism."

40. Rhyne, *Some Southern Cotton Mill Workers and Their Villages;* Norris, "Industrial Paternalist Capitalism and Local Labour Markets."

41. Scranton, "Varieties of Paternalism."

42. Consider the respective voting rates of owners and workers. The U.S. census maintained data on the percentage of farmworkers voting in national elections from 1964 to 1996. On average, farm owners and managers voted at twice the rate of foremen and laborers, although during presidential election years, the average ratio was 1.8 to 1. See U.S. Census Bureau, *Population Characteristic (P20) Reports and Detailed Tables* at www.census.gov/hhes/www/socdemo/voting/publications/p20/.

43. Scranton, "Varieties of Paternalism," 239.

44. Bittman, "Everyone Eats There."

45. Johnstone, "Old Ideals versus New Ideas in Farm Life," 150.

46. Dyer Jr., *Cultural Change in Family Firms.*

47. Johnstone, "Old Ideals versus New Ideas in Farm Life," 150. It is also worth noting that at least one author has argued that paternalism is more apt to occur in rural settings where the employer is personally known to the workers. Norris, "Industrial Paternalist Capitalism and Local Labour Markets."

48. Habraken, *The Structure of the Ordinary.*

49. Massey and Denton, *American Apartheid;* Oliver and Shapiro, *Black Wealth/White Wealth.*

50. Lauman and House, "Living Room Styles and Social Attributes"; Pratt, "The House as an Expression of Social Worlds"; Markus, *Buildings and Power;* Duncan, "Landscape Taste as a Symbol of Group Identity."

51. Duncan, "Landscape Taste as a Symbol of Group Identity."

52. Hunter, "The Symbolic Ecology of Suburbia."

53. Tilly, *Durable Inequality,* 10.

54. Knights and McCabe, "'A Different World.'"

55. Newby et al., *Property, Paternalism and Power.*

56. Newby, "The Deferential Dialectic."

57. In *Migrant,* their 1971 ethnography of New York labor camps, William H. Friedland and Dorothy Nelkin concluded that the African American workers who followed the migrant stream to New York in the late 1960s were "non-practicing citizens." Despite their U.S. citizenship, they were denied access to social and political incorporation in much the same way same that today's workers are. To understand how New York growers were able to exploit citizen workers—southern black migrants—for so long, see the work of Friedland and Nelkin: Friedland and Nelkin, *Migrant;* Friedland, "Labor Waste in New York"; Friedland, *Manufacturing Green Gold;* Nelkin, "Response to Marginality"; Nelkin, *On the Season.* Friedland and Nelkin's study of fourteen migrant camps highlights many of the same themes that my interviewees experienced: extreme precariousness of their livelihoods, apathy about their situations, and fatalism regarding the systems that had essentially trapped them in this labor market. *Migrant,* 2.

58. Jensen, *New Immigrant Settlements in Rural America;* Kandel and Cromartie, *New Patters of Hispanic Settlement in Rural America;* Lichter and Johnson, "Emerging Rural Settlement Patterns and the Geographic Redistribution of America's New Immigrants."

59. Lichter, "Immigration and the New Racial Diversity in Rural America," 16, n. 6.

60. Donato et al., "Recent Immigrant Settlement in the Nonmetropolitan United States"; Farmer and Moon, "An Empirical Examination of Characteristics of Mexican Migrants to Metropolitan and Nonmetropolitan Areas of the United States"; Kochhar, Suro, and Tafoya, *The New Latino South.*

61. Lichter, "Immigration and the New Racial Diversity in Rural America."

62. Jayaraman, *The Hands That Feed Us,* 34.

63. Even daily tasks could be a challenge, and several interviewees expressed discomfort with shopping right after work, when their clothes were soiled from the fields. They assumed that others would be offended by their appearance and look askance at them. Even perfunctory transactions at the bank could be fraught, as tellers insisted on seeing a driver's license for identification. As a result, workers routinely asked others, including employers, to cash their checks in order to avoid such interactions.

64. Lyson, *Civic Agriculture;* Berry, *The Unsettling of America;* Berry, "The Whole Horse."

65. Lichter, "Immigration and the New Racial Diversity in Rural America."

66. Three-quarters of all workers and 89 percent of non–guest workers (i.e., undocumented, resident, and citizen workers) worked in the Hudson Valley with family and/or community members from their home countries. Sixty percent worked with family and 42 percent with members of what was, in effect, a portable community. As one worker put it, "Almost my whole town is here."

67. Pfeffer and Parra, *Immigrants and the Community.* For centuries, migrant workers returned home every year after traveling for work. See Foner, "What's New about Transnationalism?"; Sassen, *Guests and Aliens;* Wyman, "Return

Migration—Old Story, New Story." This was also true of Mexican farmworkers in the United States who traditionally engaged in such circular migration. It is now clear that post-9/11 border tightening has made this more difficult. Today's undocumented workers are effectively trapped inside the United States due to increasing border controls, and thus they are more likely to stay for many years or to settle permanently. See Durand, Massey, and Zenteno, "Mexican Immigration to the United States."

68. This is not a new phenomenon. In the nineteenth and early twentieth centuries, many immigrant "birds of passage" intended to return home with U.S.-earned wages to improve the lives of their families. In the case of Italian and Scandinavian workers, for example, this was often cited as a reason for their reluctance to join trade unions. In turn, many unions became hostile to such immigrants, whose single-minded interest in earning money meant they were willing to work long hours and endure substandard working conditions, thus making it more difficult for other workers to improve their situations. See Piore, *Birds of Passage*. More recently, in the 1960s, farmworkers in Texas who lived on the Mexico-U.S. border and returned each night to Mexico were similarly unsympathetic to the UFW cause. See Schauer and Tyler, "The Unionization of Farm Labor," 16.

69. Among U.S. food workers, farmworkers were found to have the lowest potential for career advancement. See Jayaraman, *The Hands That Feed Us,* 38.

70. Workers had, on average, a sixth-grade education, and although 80 percent reported they could read and write, I observed that few could read well even in their native languages.

3. THE FARMERS

1. Grady, *The State of Agriculture in the Hudson Valley.*

2. Despite the fact that many Hudson Valley farms are conventional—a dirty word for some foodies—the owners I interviewed insisted they used as little pesticide as possible, if only to save on cost. "We measure it exactly to use the least amount, and we spray on the best days so it doesn't spread but goes where we want it," detailed one of my interviewees. "We use less than homeowners. We are really careful because it is very expensive, and we know how to save money." Hence the popularity of Integrated Pest Management (IPM), a system dedicated to using the minimum volume of chemicals necessary. It was not uncommon, I was told, to meet educated food consumers at farmers' markets in Manhattan and Westchester who wanted to discuss the pros and cons of IPM. They were described to me as clients who understand how difficult and painstaking it is to grow produce, and especially apples, without pesticides in a region with such formidable pest and fungal challenges.

3. According to the 2007 Census of Agriculture, 15 percent of the farms in the six counties in this study earned more than $100,000 and 22 percent earned more than $50,000. United States Department of Agriculture, *2007 Census of Agriculture.*

4. Hoppe et al., "Structure and Finances of U.S. Farms."

5. Daly, *Agriculture in Transition;* Ferguson, *Saving Working Landscapes;* Grady, *The State of Agriculture in the Hudson Valley;* Stanton, *The Changing Landscape of New York Agriculture in the Twentieth Century.*

6. Johnstone, "Old Ideals versus New Ideas in Farm Life," 141.

7. Hofstadter, *The Age of Reform,* 37.

8. Johnstone, "Old Ideals versus New Ideas in Farm Life," 122.

9. Beus and Dunlap, "Endorsement of Agrarian Ideology and Adherence to Agricultural Paradigms"; Dalecki and Coughenour, "Agrarianism in American Society"; Hofstadter, *The Age of Reform;* Johnstone, "Old Ideals versus New Ideas in Farm Life."

10. Quotes from Amy Courtney of Freewheelin' Farm in Davenport, California, and Brock Fulmer of the Bay Area Meat CSA in Berkeley, California, in the film *The Greenhorns* (2010), directed by Severine von Tscharner Fleming. See www.thegreenhorns.net/.

11. Writing in the 1930s, Johnstone described the misperception that the farm was "a gentle haven from the world's strife" considering that modern agriculture embraced commercialization and mechanization. Johnstone, "Old Ideals versus New Ideas in Farm Life," 165–66. And, as Hofstadter pointed out almost two decades later, a U.S. farmer was someone whose self-sufficiency "was usually forced upon him by lack of transportation or markets." Hofstadter, *The Age of Reform,* 23. Despite this disconnect, farmers have long benefited from tapping into the sentimental characterizations of their kind. More recently, artisanship was revalorized in books such as *Shop Class and Soul Craft: An Inquiry into the Value of Work,* the best seller by academic-turned-mechanic Matthew B. Crawford, and *The Craftsman,* by sociologist Richard Sennet.

12. Johnson, "Looking for Your Own Dream Farm?"; Russell, "Small Farm Dreams."

13. Examples of books that detail the trials and transformations of new farmers include Friend, *Hit by a Farm* (urbanite turns sheep farmer when her partner decides to follow a dream); McCorkindale, *Confessions of a Counterfeit Farm Girl* (East Coaster gives up a six-figure job to move south and run a cattle farm); Kimball, *The Dirty Life* (heels and manicures in Manhattan are traded for Carhartt boots, a Leatherman, and animal evisceration in New York's North Country); and Kilmer-Purcell, *The Bucolic Plague* (a former drag queen and his partner restore a Hudson Valley mansion and its farmland). It is worth noting that these memoirs were penned primarily by women, including lesbians, and a gay male, groups that are not at all representative of the industry.

14. Rehak, "Growing Pains."

15. Mishra et al., *Income, Wealth, and Economic Well-Being of Farm Households.*

16. Low and Vogel, *Direct and Intermediate Marketing of Local Foods in the United States,* 2; Cloud, "Eating Better Than Organic"; Martin, "Is a Food Revolution Now in Season?"; Matthews, "In Hudson Valley, Farmers' Markets Offer Local Food and Local Chat"; Severson, "Eating: Why Roots Matter More"; Zawisza, *FDA Announces Findings from Investigation of Foodborne E. Coli O157.*

See also regional magazines that promote local farmers and the markets and restaurants that serve their goods such as *The Valley Table* (www.valleytable.com) and *Edible Hudson Valley* (www.ediblecommunities.com/hudsonvalley).

17. DuPuis ("Angels and Vegetables," 34) explores the question of ethical eating and finds that the following authors all deal with this question: Lappé and Terry, *Grub;* Nestle, *What to Eat;* Pollan, *The Omnivore's Dilemma;* Singer, *The Ethics of What We Eat;* Tuttle, *The World Peace Diet.*

18. See, for example, Neuman and Barboza, "U.S. Drops Inspector of Food in China." An ABC TV segment titled "Whole Foods Market China Organic California Blend Boycott Products of China," produced by the Chicago affiliate WJLA TV's I-Team Division, has also made regular rounds on the Internet since it was aired in 2008. See www.youtube.com/watch?v=JQ31Ljd9T_Y&feature=youtu.be. The corresponding list of Whole Foods organic produce from China is at www.acc-tv.com/images/wjla/news/iteamwholefoodslist052108.pdf, and Whole Foods responded to the story at www.wholefoodsmarket.com/blog/whole-story/organic-china-possible.

19. Sander, "Source of Deadly E. Coli Is Found"; Zawisza, *FDA Announces Findings from Investigation of Foodborne E. Coli O157.*

20. The claim that food traveled an average of 1,500 miles from farm to plate was offered by Pirog et al., who examined previous studies of food mileage. See Pirog et al., *Food, Fuel, and Freeways.*

21. The blog eatlocalchallenge.com offers an overview of the reasons why one should eat locally and how.

22. Corderi, "Should We Irradiate Fruits and Vegetables?"

23. Shteyngart, "Escape to New York's Hudson Valley."

24. Johnson, "Garden of Eating."

25. Hofstadter, *The Age of Reform,* 24.

26. Barritt, "Food for Thought."

27. *The Valley Table,* one Hudson Valley–based food and farms magazine, puts the count of farmers' markets in the six counties involved in my study at forty-six (www.valleytable.com/farms.php).

28. Low and Vogel, *Direct and Intermediate Marketing of Local Foods in the United States.*

29. Johnstone, "Old Ideals versus New Ideas in Farm Life," 142.

30. Research has shown farmers are open to consumer input on chemical use (Hunt, "Consumer Interactions and Influences on Farmers' Market Vendors") and that local food systems can promote trust, which increases consumer demand (Moser, Raefelli, and Thilmany, "Identifying Influential Attributes in WTP for Agro-Food with Credence Based Public Good Attributes").

31. Pollan, *The Omnivore's Dilemma,* 240.

32. Matthews, "In Hudson Valley, Farmers' Markets Offer Local Food and Local Chat."

33. Low and Vogel, *Direct and Intermediate Marketing of Local Foods in the United States.*

34. Clancy and Ruhf, "Is Local Enough?"; Lev and Gwin, "Filling in the Gaps."

35. Low and Vogel, *Direct and Intermediate Marketing of Local Foods in the United States.*

36. Grady, *The State of Agriculture in the Hudson Valley,* 20.

37. Pollan, "Vote for the Dinner Party."

38. On farmers' markets, see Oberholtzer and Grow, *Producer-only Farmers' Markets in the Mid-Atlantic Region.* On CSAs, see Jacques and Collins, "Community Supported Agriculture"; Schnell, "Food with a Farmer's Face"; Feagan and Henderson, "Devon Acres CSA." On local food systems, see Ostrom, "Everyday Meanings of 'Local Food'"; Hinrichs "Embeddedness and Local Food Systems." For an explanation of civic agriculture and how it promotes community-based agriculture and social capital, see Lyson, *Civic Agriculture;* Lyson and Green, "The Agricultural Marketscape." For how local food systems promote democracy, see Feenstra, "Local Food Systems and Sustainable Communities."

39. Ferguson, *Saving Working Landscapes,* 5.

40. New York Agriculture and Markets Law Article 25 AA.

41. New York Agriculture and Markets Law § 308.

42. New York Agriculture and Markets Law § 303, 306, 308 (3).

43. Martin, *Illegal Immigration and the Colonization of the American Labor Market.*

44. Bucholz, *Farmworker Housing in New York State;* Hamilton, "Farmworker Housing in New York State."

45. Westneat, "The Fruits of Our Labor Absurdity." See also Johnson, "Farmers Strain to Hire American Workers in Place of Migrant Labor." In addition, the state of Georgia conducted an unintended experiment in 2011, when highly restrictive state immigration laws drove many Latinos out of the state and farms lost 40 percent of their work force. Farmers were unable to replace them even a year later. See Powell, "The Law Of Unintended Consequences."

46. Rivers, *Farm Hands.*

47. Piore, *Birds of Passage;* Reich, Gordon, and Edwards, "A Theory of Labor Market Segmentation."

48. Piore, *Birds of Passage.*

49. Stewart, *It's a Long Road to a Tomato,* 23.

50. Bernstein, "Immigrants Go from Farms to Jails, and a Climate of Fear Settles In."

51. Baker, "Agricultural Labor."

52. Bernstein, "Immigrants Go from Farms to Jails, and a Climate of Fear Settles In."

53. Suarez, "Memo Regarding Farmworker Advocacy Day."

54. Sherwood, "The New Migrants."

55. Alejandro Portes used this comparison at a presentation during the conference Second Cumbre of the Great Plains: Re-visioning Latino America—New Perspectives on Migration, Transnationalism, and Integration, at the University of Omaha, April 22–24, 2005.

56. U.S. President's Commission on Migratory Labor, *Migratory Labor in American Agriculture*.

57. Pfeffer and Parra, *Immigrants and the Community*.

58. One industry resource I ordered, "Spanish for Greenhouse Operators," was well intentioned, but the recording left no pauses between the vocabulary words and sped by so quickly that it was not very useful.

4. SUSTAINABLE JOBS?

1. Pfeffer and Parra, *Immigrants and the Community*.

2. Griffith, *American Guestworkers*, 62; Margolies, *Training Needs Assessment of Farm Workers in Orange and Sullivan Counties, NY*, 10–11; North and Holt, *The Apple Industry Work Force*, 393, 395; Pfeffer and Parra, *Immigrants and the Community: Former Farmworkers*, 4. Substantial scholarly literature addresses racial discrimination against blacks socially and in the workplace. See Bertrand and Mullainathan, "Are Emily and Greg More Employable Than Lakisha and Jamal?"; Bonilla-Silva, *Racism Without Racists;* Brown et al., *White-Washing Race;* Massey and Denton, *American Apartheid;* Pager, "The Mark of a Criminal Record"; Polikoff, "Racial Inequality and the Black Ghetto"; Wilson, *When Work Disappears.* The displacement of low-wage African American workers by immigrants has received a bit more attention in the past few years. See, for example, Stephen Steinberg's essays arguing that low-wage blacks have been replaced by immigrants (Steinberg, "Immigration, African Americans, and Race Discourse"; Steinberg, "Immigration, African Americans, and Race Discourse") and rejoinders by several academics in *New Politics* 10, no. 4 (Winter 2006) and *New Labor Forum* 15, no. 1 (Spring 2006).

3. Barr, *Liberalism to the Test*, 5; Nelkin, *On the Season*, 3.

4. Job opportunities grew in Florida, both in the construction industry and for the technology-extended orange-growing season.

5. Analyzing the labor profiles of specific farms gives an in-depth sense of the change in the work force, but the farms' hiring decisions have to be seen in a structural context. Hiring processes are best understood by examining an entire industry or regional labor market as opposed to individual employers. See Tilly, *Durable Inequality*. In this way we can understand how a series of individual choices is institutionalized as a norm throughout the entire region.

6. Neoclassical economic interpretations of international migration relate such changes to the rational motivation of immigrants who engage in individual cost-benefit analysis to make the decision to emigrate. See Borjas, "Economic Theory and International Migration"; Galbraith, *The Nature of Mass Poverty;* Greenwood, "Research on Internal Migration in the United States"; Sjaastad, "The Costs and Returns of Human Migration"; Todaro, *Internal Migration in Developing Countries*. This approach has been criticized for ignoring large-scale migration that is occasioned by larger structural variables such as land reform, neoliberal policies, or

climate change. See Arizpe, "The Rural Exodus in Mexico and Mexican Migration to the United States"; Papademetriou and Hopple, "Causal Modeling in International Migration Research"; Portes and Watson, *Labor Class and the International System;* Sassen, *The Mobility of Labor and Capital.* The late-twentieth-century reconfiguration of agriculture in Mexico away from communal landownership arrangements was one such factor, since it engendered a shift from subsistence to commodity farming, the consolidation of landholdings, and the state's retraction of aid to farmers. These changes, some of which stemmed from free trade agreements with the United States, created a class of displaced farmers who were ripe for migration to the United States, many of whom ended up in the new immigrant destinations. See Hatton and Williamson, "What Drove the Mass Migrations from Europe in the Late Nineteenth Century?"; Massey, "International Migration and Economic Development in Comparative Perspective."

7. Dual labor market theory provides a demand-side perspective that analyzes the division of jobs in industrialized economies into primary and secondary sectors.

8. Massey et al., *Return to Aztlan;* Portes, *Latin Journey;* Zabin, *Mixtec Migrants in California Agriculture.* Network migration theory is the most prominent perspective on increasing immigration. It bridges both demand and supply approaches, and its adherents posit that reciprocal social relationships, rooted in the sending country's hometowns, promote migration to receiving countries. As such, pioneering immigrants to a regional economy not only serve as recruiting agents for others in their hometowns, but they also offer necessary social and economic support to new immigrants.

9. Krissam, "Sin Coyote Ni Patrón."

10. For an analysis and critique of the migration network model of immigration, see Krissam, "Sin Coyote Ni Patrón." Krissam argues that network migration theory glosses over the role of employers, labor smugglers, and others. Furthermore, he posits that network migration theory does not fully explain the continued expansion of immigrants to a particular area. One major problem he identifies is that without a comprehensive explanation for expanding immigration, policy makers cannot appropriately address immigration concerns. And, as a result, it provides "cover" for those demand-side actors who are responsible for increasing immigration. In response, Krissam offers his own model for explaining increasing immigration.

11. Krissam, "Sin Coyote Ni Patrón." Although a few studies do address employers' roles (Griffith, *American Guestworkers;* Izcara Palacios, "La adicción a la mano de obra ilegal"; Krissam, "Sin Coyote Ni Patrón"; Piore, *Birds of Passage*), analyses of the hiring process more commonly look at how the "migration service industry" aids not only employers (Zúñiga and Hernández-León, "Introduction," xvii), but also the *coyotes,* immigrant entrepreneurs, and labor contractors who facilitate job placement. See Hernández-León, *Metropolitan Migrants;* Mahler, *American Dreaming.* Moreover, immigration literature generally fails to address employers' preference for one ethnic or racial group over another or their role in labor market ethnic succession. Exceptions include Green, "Women and Immigrants in the Sweatshop"; Griffith, *The Estuary's Gift;* Griffith, *American Guestworkers.*

12. Millard and Chapa, *Apple Pie and Enchiladas;* Murphy, Blanchard, and Hill, *Latino Workers in the Contemporary South;* Zúñiga and Hernández-León, *New Destinations of Mexican Immigration in the United States.*

13. Massey, "The New Geography of Mexican Immigration."

14. Rytina, *IRCA Legalization Effects.*

15. Griffith, *American Guestworkers.*

16. This number is from the New York State Employment Service. Nelkin, *On the Season,* 3. A 1959 demographic study of New York's black migrants revealed youth working on the state's farms: 18 percent were under the age of fourteen and 15 percent were between the ages of fifteen and nineteen. Larson, *Migratory Agricultural Workers in the Eastern Seaboard States, Rural Poverty in the United States,* quoted in Nelkin, *On the Season,* 4.

17. In my interviews with Hudson Valley farmworkers and farmers, I heard of only two African American workers who had been employed in the last decade, but I never encountered any. Of the several hundred workers I identified on the farms where I conducted interviews, not one was African American. I did, however, meet many former African American farmworkers around New York, including many who described working as entire families in the fields. Several African American service providers who worked for nonprofits catering to the state's farmworkers, were former farmworkers themselves.

18. My reference here to Latin Americans and Latinos does not include Puerto Ricans, who are U.S. citizens. Puerto Ricans did staff New York farms, and canneries in particular, beginning in the 1940s, though in comparatively small numbers compared to African American workers.

19. The works of David Griffith and Cindy Hahamovitch offer insight into the development and history of West Indian—primarily Jamaican—farmworkers in East Coast agriculture. See, for example, Griffith, *American Guestworkers;* Hahamovitch, *The Fruits of Their Labor.*

20. Gray, *The Hudson Valley Farmworker Report,* 17.

21. Maloney and Bills, *The New York State Agricultural Immigration and Human Resource Management Issues Study.*

22. Calavita, *Inside the State.*

23. In the 1990s most U.S. agricultural guest workers were from Mexico, and Mexican guest workers became more common on New York farms. Hahamovitch provides an analysis of the inception of this program. Hahamovitch, *The Fruits of Their Labor.* Griffith *(American Guestworkers)* offers insight on the development of the program.

24. North and Holt, *The Apple Industry Work Force;* Pfeffer and Parra, *Immigrants and the Community: Former Farmworkers.*

25. Lapp, "Managing Migration."

26. Puerto Ricans had been brought to the United States for agricultural work beginning in the early 1900s, but it was not until the 1940s that their recruitment became more official through a joint program of the Puerto Rican and U.S. Departments of Labor. See p. 164 n. 110.

27. García-Colón, "Claiming Equality," 283.

28. Murphy, Blanchard, and Hill, *Latino Workers in the Contemporary South;* Zúñiga and Hernández-León, *New Destinations of Mexican Immigration in the United States.*

29. Haslip-Viera and Baver, *Latinos in New York.*

30. Smith, "Mexicans in New York."

31. For a study of Oaxacan immigrants to Poughkeepsie, see Mountz and Wright, "Daily Life in the Transnational Migrant Community of San Agustin, Oaxaca, and Poughkeepsie, New York." For a documentary film on transnational migrants, see Rivera, *The Sixth Section.* Using digital animation and interview footage, Rivera documents the efforts by a hometown association of immigrant men living in Newburgh, New York, to conduct public works in their hometown of Boquerón, Mexico. In part, he examines the power they have to alter politics at home.

32. Zúñiga and Hernández-León, "A New Destination of an Old Migration," 126. See also Durand, Massey, and Parrado, "The New Era of Mexican Migration to the United States"; Neuman and Tienda, "The Settlement and Secondary Migration Patterns of Legalized Immigrants."

33. Griffith, *American Guestworkers.*

34. Lehmann, *Migrant Farmworkers of Wayne County, New York,* 1.

35. Griffith, *American Guestworkers,* 62.

36. Phillips and Massey, "The New Labor Market"; Rivera-Batiz, "Undocumented Workers in the Labor Market."

37. Martin, "The Endless Debate."

38. It is worth noting in this regard that guest worker programs serve as a sort of background check since workers must be approved in their home countries for the positions. Such checks include physicals and other measures.

39. Griffith, *American Guestworkers,* 167.

40. Bayne, *County at Large;* Horne, "Life on a Rocky Farm."

41. Hahamovitch, *The Fruits of Their Labor.*

42. Piore, *Birds of Passage;* Reich, Gordon, and Edwards, "A Theory of Labor Market Segmentation."

43. Fisher, *The Harvest Labor Market in California.*

44. Barr, *Liberalism to the Test.*

45. Griffith, *American Guestworkers,* 9.

46. Kelley, *Race Rebels.*

47. Moss and Tilly, *Stories Employers Tell;* Steinberg, "Immigration, African Americans, and Race Discourse," 2005; Waldinger and Lichter, *How the Other Half Works.*

48. Green, "Women and Immigrants in the Sweatshop."

49. Adams, *Farm Management,* 519–25.

50. Maldonado, "Racial Triangulation of Latino/a Workers by Agricultural Employers."

51. In 2005, for example, Cornell Cooperative Extension of Jefferson County, in association with Cornell University's Department of Applied Economics and

Management, organized the North Country Hispanic Workforce Conference, advertised as "targeted for owners or managers of farms that hire Hispanic employees, as well as farmers who are considering hiring Hispanic Employees." In addition, in 2005, Cornell Cooperative Extension organized the "Managing the Hispanic Workforce" Conference in Syracuse, the proceedings of which were available for purchase. Thomas R. Maloney, Senior Extension Associate at Cornell University, in particular, conducts and publishes research on farm labor management issues specific to New York.

52. See, for example, Morales, "How to Make the Most of My Multicultural Workforce"; Nevaer and Ekstein, *HR and the New Hispanic Workforce;* Maloney, "Understanding the Dimensions of Culture."

53. Sherwood, "The New Migrants."

54. Johnston, "Farm Worker Killed on Job."

5. TOWARD A COMPREHENSIVE FOOD ETHIC

1. Bobrow-Strain, "Kills a Body Twelve Ways."

2. Pollan, "Vote for the Dinner Party."

3. Ibid.

4. Bittman, "Everyone Eats There."

5. The initial versions of the bill were A1877 A and S2247 A; they were amended in committee as A1867 B and S2247, and the versions that went to floor votes in 2010 were A11569 and S8223. For a detailed analysis of the proposed legislation, see Telega and Maloney, "Provisions and Politics of the Farmworker Fair Labor Practices Act," and for a summary and analysis of the bill's path through the New York State Assembly, see Telega and Maloney, *Legislative Actions on Overtime Pay and Collective Bargaining.*

6. Edid, *Farm Labor Organizing;* Ganz, *Why David Sometimes Wins;* Rothenberg, *With These Hands.*

7. Millones, "Kennedy and Javits Are Shocked by Housing of Migrants Upstate."

8. Friedland and Nelkin, *Migrant;* Nelkin, *On the Season.*

9. Edid, *Agricultural Labor Markets in New York State and Implications for Labor Policy.*

10. Mendez and Diaz, *Separate & Unequal.*

11. Farmworker organizing is and historically has been extremely difficult. See Adams, "Quiescence Despite Privation"; Compa, *Unfair Advantage;* Edid, *Farm Labor Organizing;* Ganz, *Why David Sometimes Wins;* Majka and Majka, *Farm Workers, Agribusiness, and the State;* Majka and Majka, "Organizing U.S. Farm Workers"; Morin, *The Organizability of Farm Labor in the United States;* Sánchez and Romo, *Organizing Mexican Undocumented Farm Workers on Both Sides of the Border;* Schauer and Tyler, "The Unionization of Farm Labor." Organizing has a long history of being squashed by growers. See Goldfarb, *Migrant Farm Workers;* Grossardt, "Harvest(ing) Hoboes"; Hahamovitch, *The Fruits of Their Labor;* Majka

and Majka, *Farm Workers, Agribusiness, and the State;* Meister and Loftis, *A Long Time Coming.*

12. The New York legislative structure gives significant control to the Senate Majority Leader and Assembly Speaker, including total control over the legislative calendar and whether a bill that has been approved by a committee will be voted on by the full chamber. In addition, the parties maintain tight control over members' voting decisions through a system of rewards and punishments, such as committee assignments and chair positions. This results in bills reaching the floor for a vote only when the leadership supports them and passage is almost guaranteed. Creelan and Moulton, *The New York State Legislative Process;* Norden, Pozen, and Foster, *Unfinished Business;* Stengel, Norden, and Seago, *Still Broken.*

13. New York State Consolidated Laws, Chapter 31 Labor Laws: Article 7 §212, "Drinking water for farm laborers"; Article 7, General Provisions §212-D, "Field sanitation for farm hand workers, farm field workers and farm food processing workers"; Article 19-A, "Minimum wage standards and protective labor practices for farm workers" §673[2], Minimum wage, and §674, Regulations.

14. State laws differ from federal ones, including Occupational Safety and Health Administration (OSHA) laws. Employers are required to follow all laws and adhere to the stricter standard. The New York State law is stricter than the federal standard, which requires drinking water and sanitation with hand-washing facilities only when eleven or more workers are in the fields. See OSHA Regulations (Standards—29 CFR) Part 1928, Occupational Safety and Health Standards for Agriculture, Standard 1928.110, Field Sanitation, or Fact Sheet No. OSHA 92–25 at www.nmsu.edu/safety/resources/forms/OSHA-FieldSanitationStd1992.pdf.

15. Stigler, "The Theory of Economic Regulation."

16. United States Department of Agriculture, National Agricultural Statistics Service, *Table 1.*

17. The New York Farm Bureau contributes to national lobbying and also does sophisticated statewide lobbying that includes professional lobbyists, growers, and an e-lobby system. To say that the Farm Bureau network nationwide has been active in agricultural policy decisions would be an understatement. In fact, the New York Farm Bureau was long believed to be an arm of government, a perception that had lasting credence with the public. Other farm business organizations that lobbied against the Farmworkers Fair Labor Practices Act include the New York Horticultural Society and Agricultural Affiliates.

18. Legal investigations constitute another form of harassment. Farmworker Legal Services of New York (FLSNY) has endured legislators' scrutiny after complaints were lodged with members of Congress questioning its use of federal funds. On each occasion FLSNY was cleared. In 2006, the Mid-Hudson Migrant Education Outreach Program was investigated by the United States Department of Education Office of Inspector General due to allegations that it had overstepped its bounds by lobbying. The audit showed no lobbying was done by the program staff.

19. Nieves, "Farm Workers Stand Tough and Walk Out."

20. Moore, *The Slaves We Rent.*

21. Majka and Majka, *Farm Workers, Agribusiness, and the State;* McWilliams, *Factories in the Field.*

22. Tomasek, "The Migrant Problem and Pressure Group Politics."

23. Hahamovitch, *The Fruits of Their Labor.*

24. Hightower, *Hard Tomatoes, Hard Times.*

25. According to the internal review committee's report, "Based on statements made to the committee, opponents are intervening with Cornell Administration, raising their concerns with farm employer organizations such as the New York Farm Bureau and the state Horticultural Society, to challenge state support for Cornell and challenging its tax status through the Internal Review Service." Review Committee, *Internal Review of the Cornell Migrant Program,* 14.

26. The former director stayed on until October 2007 to write a history of the CMP and archive the organization's materials.

27. Neil Schwartzbach of the Cornell Community and Rural Development Institute conducted forty interviews for an oral history of the CMP.

28. The following entered into a "covenant" with the Rural and Migrant Ministry: the Capital Area Baptist Association, the Mid-Hudson Baptist Association, the Presbytery of Albany, the Presbytery of the Hudson River, the New York Annual Conference of the United Methodist Church, the Regional Synod of New York—Reformed Church in America, the Episcopal Diocese of New York, the Roman Catholic Diocese of Albany, the Roman Catholic Diocese of Rochester, the New York Yearly Meeting of the Society of Friends (Quakers), the New York Conference of the United Church of Christ, the Episcopal Diocese of Rochester, and the New York State Association of American Baptists.

29. "New York Senate Legislative Resolution honoring the many contribution of New York State farmworkers and recognizing their need for fair labor practices," No. 92113–01–9, introduced by Senator Bill Perkins, April 29, 2009.

30. Kennedy, "This Must Be New York's Final Harvest of Shame."

31. Matteson, "Interview with Julie Suarez."

32. Norton, "Don't Turn Farms into Factories," 23; New York Farm Bureau, *Farmworker Bill Defeated;* The Assembly Minority Conference, *Assembly Minority.* See also Hawley, *Hawley Stands Opposed to Farm Death Bill.* Language about the bill killing agriculture in New York State peppered the comments of state senators opposed to the bill during the senate's extraordinary session on August 3, 2010, when the bill went to a vote.

33. See the video "Fact vs. Myth on Farm Worker Legislation," available at www.nysenate.gov/video/2009/dec/14/facts-vs-myth-farm-worker-legislation. Describing farm owners as farmworkers is not a new idea. In 1951 C. Vann Woodward described the phenomenon, writing, "The landowner was so poor and distressed that he forgot that he was a capitalist ... so weary of hand and sick of spirit that he imagined himself in precisely the same plight as the hired man." In *Origins of the New South* (Baton Rouge, 1951), cited in Hofstadter, *The Age of Reform,* 47.

34. One of its co-sponsors voted no, as did an Espada ally, who was vocal in his support during the March 2010 meeting between the JFW and the Democratic

conference leader. A third vote was lost because a senator who had vocally supported the bill left the chamber.

35. Confessore and Hartocollis, "Albany Approves No-Fault Divorce and Domestic Workers' Rights."

36. Guthman, *Agrarian Dreams,* 177.

37. United States Department of Agriculture data for October 7–13, 2012, reports that fieldworkers in the Northeast I Region, in which New York is included, earned $10.99 an hour, on average. See http://usda01.library.cornell.edu/usda/nass /FarmLabo//2010s/2012/FarmLabo-11-19-2012.pdf.

38. Martin, "Calculating the Costs and Benefits."

39. Brown, "The New Geopolitics of Food"; Bloom, *American Wasteland.*

40. www.swantonberryfarm.com/.

BIBLIOGRAPHY

Adams, Jane. "Quiescence Despite Privation: Explaining the Absence of a Farm Laborers' Movement in Southern Illinois." *Comparative Studies in Society and History* 39, no. 3 (1997): 550–71.

Adams, Richard Laban. *Farm Management: A Text-book for Student, Investigator, and Investor.* New York: McGraw-Hill Book Publishing, 1921.

Allen, Patricia. *The Human Face of Sustainable Agriculture.* Santa Cruz, CA: University of California Santa Cruz Center for Agroecology & Sustainable Food Systems, November 1994.

———. "Mining for Justice in the Food System: Perceptions, Practices, and Possibilities." *Agriculture and Human Values* 25, no. 2 (2008): 157–61.

———. *Together at the Table: Sustainability and Sustenance in the American Agrifood System.* University Park, PA: Pennsylvania State University Press, 2004.

Allen, Patricia, Margaret FitzSimmons, Michael Goodman, and Keith Warner. "Shifting Plates in the Agrifood Landscape: The Tectonics of Alternative Agrifood Initiatives in California." *Journal of Rural Studies* 19, no. 1 (2003): 61–75.

Alston, Lee J., and Joseph P. Ferrie. *Southern Paternalism and the American Welfare State: Economics, Politics, and Institutions in the South.* Cambridge: Cambridge University Press, 1999.

Alteri, Miguel. *Agroecology: The Science of Sustainable Agriculture.* Boulder, CO: Westview Press, 1995.

Amidon, Beulah. *What's Next for New York's Joads?* Consumers League of New York, September 1946.

Arizpe, Lourdes. "The Rural Exodus in Mexico and Mexican Migration to the United States." *International Migration Review* 15, no. 4 (1982): 626–49.

Assembly Minority Conference. *Assembly Minority: So-called "Farmworkers Fair Labor Practices Act" Will Force More Family Farms Out of Business.* Albany: New York State Assembly, June 9, 2009. http://assembly.state.ny.us/Minority /20090609/.

Back, Adina. *Student Voices from World War II and the McCarthy Era.* New York: Brooklyn College, City University of New York, John Jay College, CUNY and

American Social History Project / Center for Media and Learning at the Graduate Center, City University of New York. www.ashp.cuny.edu/oralhistory/index.html.

Baker, Paul. "Agricultural Labor." *New York Fruit Quarterly* 14, no. 2 (2006).

Barr, Donald. *Liberalism to the Test: African-American Migrant Farmworkers and the State of New York.* New York: State University of New York, New York State African American Institute, February 1988.

Barritt, T. W. "Food for Thought: A Sense of Place." *Edible Hudson Valley,* 2012. www.ediblecommunities.com/hudsonvalley/spring-2012/food-for-thought.htm.

Bayne, Martha Collin. *County at Large.* Poughkeepsie, NY: The Women's City and County Club with Vassar College, 1937.

Belasco, Warren. *Appetite for Change: How the Counterculture Took on the Food Industry 1966–1988.* New York: Pantheon, 1989.

Bennett, John W. "Paternalism." *International Encyclopedia of the Social Sciences.* New York: Macmillan, 1968.

Bernstein, Nina. "Immigrants Go from Farms to Jails, and a Climate of Fear Settles In." *New York Times,* December 24, 2006. www.nytimes.com/2006/12/24/nyregion/24migrant.html?pagewanted=all.

Berry, Wendell. *The Unsettling of America: Culture and Agriculture.* San Francisco: Sierra Club Books, 1977.

———. *What Are People For?* New York: North Point Press, 1990.

———. "The Whole Horse." In *The Art of the Common-place: The Agrarian Essays of Wendell Berry,* edited by Norman Wirzba, 236–48. Washington, DC: Counterpoint, 2002.

Bertrand, Marianne, and Sendhil Mullainathan. "Are Emily and Greg More Employable Than Lakisha and Jamal? A Field Experiment on Labor Market Discrimination." *American Economic Review* 94, no. 4 (2004): 991–1013.

Beus, Curtis, and Riley E. Dunlap. "Endorsement of Agrarian Ideology and Adherence to Agricultural Paradigms." *Rural Sociology* 59, no. 3 (1994): 462–84.

Bittman, Mark. "Everyone Eats There." *New York Times Magazine,* October 10, 2012. www.nytimes.com/2012/10/14/magazine/californias-central-valley-land-of-a-billion-vegetables.html?pagewanted=all.

Black, Jan. "Barry Estabrook's 'Tomatoland,' an Indictment of Modern Agriculture." *Washington Post.* June 10, 2011. http://articles.washingtonpost.com/2011-06-10/entertainment/35233611_1_winter-tomato-immokalee-workers-barry-estabrook.

Bloom, Jonathan. *American Wasteland: How America Throws Away Nearly Half of Its Food (and What We Can Do About It).* Cambridge, MA: Da Capo Press, 2010.

Bobrow-Strain, Aaron. "Kills a Body Twelve Ways: Bread Fear and the Politics of 'What to Eat?'" *Gastronomica: The Journal of Food and Culture* 7, no. 3 (2007): 45–52.

Bonilla-Santiago, Gloria. "A Case Study of Puerto Rican Migrant Farmworkers Organizational Effectiveness in New Jersey." PhD diss., City University of New York, 1986.

Bonilla-Silva, Eduardo. *Racism without Racists: Color-blind Racism and the Persistence of Racial Inequality in the United States*. New York: Rowman & Littlefield Publishers, 2003.

Borjas, George J. "Economic Theory and International Migration." *International Migration Review* 23, no. 3 (1989): 457–85.

Born, Branden, and Marc Purcell. "Avoiding the Local Trap: Scale and Food Systems in Planning Research." *Journal of Planning Education & Research* 26, no. 2 (2006): 195–207.

Brewster, John M. *A Philosopher among Economists*. Philadelphia: J. T. Murphy, 1970.

Brown, Lester R. "The New Geopolitics of Food." *Foreign Policy,* May/June 2011. www.foreignpolicy.com/articles/2011/04/25/the_new_geopolitics_of_food.

Brown, Michael K., Martin Carnoy, Elliot Currie, Troy Duster, David B. Oppenheimer, Marjorie M. Shultz, and David Wellman. *White-Washing Race: The Myth of a Color-Blind Society*. Berkeley: University of California Press, 2003.

Brown, Sandy, and Christy Getz. "Privatizing Farm Worker Justice: Regulating Labor through Voluntary Certification and Labeling." *GeoForum* 39, no. 3 (2008): 1184–96.

———. "Toward Domestic Fair Trade? Farm Labor, Food Localism, and the 'Family Scale' Farm." *GeoJournal* 73 , no. 1 (2008): 11–22.

Bruegel, Martin. *Farm, Shop, Landing: The Rise of a Market Society in the Hudson Valley, 1790–1860*. Durham, NC: Duke University Press, 2002.

Bucholz, Sue. *Farmworker Housing in New York State: Analysis and Recommendations*. Rochester, NY: Rural Opportunities, October 2000.

Burch Jr., William R. *Daydreams and Nightmares: A Sociological Essay on the American Environment*. New York: Harper & Row, 1971.

Buttel, Frederick H. "Agriculture, Environment, and Social Change: Some Emergent Issues." In *The Rural Sociology of the Advanced Societies: Critical Perspectives,* edited by Frederick H. Buttel and Howard Newby, 453–88. Montclair, NJ: Allanheld, Osmun, 1980.

Buttel, Frederick H., and William L. Flinn. "Sources and Consequences of Agrarian Values in American Society." *Rural Sociology* 40, no. 2 (1975): 134–51.

Calavita, Kitty. *Inside the State: The Bracero Program, Immigration and the I.N.S.* New York: Routledge, 1992.

Carlson, Allan. "Agrarianism Reborn: On the Curious Return of the Small Family Farm." *Intercollegiate Review* 43, no. 1 (2008).

Carlson, John E., and Maurice E. McLeod. "A Comparison of Agrarianism in Washington, Idaho, and Wisconsin." *Rural Sociology* 43, no. 1 (1978): 134–51.

Carpenter, Stephanie A. *On the Farm Front: The Women's Land Army in World War II*. DeKalb, IL: Northern Illinois University Press, 2003.

Claffey, James E. "Anti-immigrant Violence in Suburbia." *Social Text* 24, no. 3 (2006).

Clancy, Kate, and Kathryn Ruhf. "Is Local Enough? Some Arguments for Regional Food Systems." *Choices Magazine* 25, no. 1 (2010). www.farmdoc.illinois.edu /policy/choices/20101/2010108/2010108.html.

Close, Kathryn. *The Joads of New York*. New York: Consumers League of New York, January 1945.

Cloud, John. "Eating Better Than Organic." *Time,* March 2, 2007. www.time.com /time/magazine/article/0,9171,1595245,00.html.

Compa, Lance. *Unfair Advantage: Workers' Freedom of Association in the United States under International Human Rights Standards.* Human Rights Watch, August 2000.

Confessore, Nicholas and Anemona Hartocollis. "Albany Approves No-Fault Divorce and Domestic Workers' Rights." *New York Times,* July 1, 2010. www.nytimes. com/2010/07/02/nyregion/02albany.html?_r=1&scp=2&sq=domestic%20 workers&st=cse.

Conford, Philip. *The Origins of the Organic Movement*. Edinburgh, Scotland: Floris Books, 2001.

Corderi, Victoria. "Should We Irradiate Fruits and Vegetables?" *Dateline NBC,* March 25, 2007. www.nbcnews.com/id/17758666/ns/dateline_nbc/#.US_vHzfg58s.

Creelan, Jeremy M., and Laura M. Moulton. *The New York State Legislative Process: An Evaluation and Blueprint for Reform*. New York: Brennan Center for Justice at New York University School of Law, 2004.

Crèvecoeur, J. Hector St. John de. *Letters from an American Farmer*. New York: E. P. Dutton & Co., 1972.

Cunningham, Douglas R. 1970. "The Non-voting Farmworker: Disenfranchisement by Design?" *University of California at Davis Law Review: Legal Problems of Agricultural Labor* 2, no. 1 (1970): 217–35.

Dalecki, Michael G., and C. Milton Coughenour. "Agrarianism in American Society." *Rural Sociology* 57, no. 1 (1992): 48–64.

Daly, Jayne. *Agriculture in Transition: Recent Trends in the Hudson Valley*. Cold Spring, NY: Glynwood Center, 2004.

Danbom, David B. "Romantic Agrarianism in Twentieth-Century America." *Agricultural History* 65, no. 4 (1991): 1–12.

Daniel, Cletus E. *Bitter Harvest: A History of California Farmworkers, 1870–1941*. Ithaca, NY: Cornell University Press, 1981.

Danielsson, Claire. *To Celebrate the Land*. New Paltz, NY: Mohonk Consultations, 2003.

Department of Labor Employment and Training Administration. *Labor Certification Process for the Temporary Employment of Aliens in Agriculture in the United States: 2012 Adverse Effect Wage Rates, Federal Registry 76 FR 79711, Document Number 2011-32842*, December 22, 2011. https://federalregister.gov/a/2011–32842.

Diamond, Adam, and Ricardo Soto. *Facts on Direct-to-Consumer Food Marketing: Incorporating Data from the 2007 Census of Agriculture*. Washington, DC: United States Department of Agriculture, Agricultural Marketing Service, 2009. www.thegreenhorns.net/resources/Facts_on_Direct_to_Consumer_ Food_Marketing.pdf.

DiNapoli, Thomas P., and Kenneth B. Bleiwas. *Farmers' Markets in New York City*. New York: Office of the State Comptroller, August 2012. www.osc.state.ny.us /osdc/farmersmarkets_rpt6–2013.pdf.

Dolan, Matthew. "New Detroit Farm Plan Taking Root." *Wall Street Journal*. July 6, 2012. online.wsj.com/article/SB10001424052702304898704577479090390757 7800.html.

Donato, Katharine M., Charles M. Tolbert II, Alfred Nucci, and Yukio Kawano. "Recent Immigrant Settlement in the Nonmetropolitan United States: Evidence from Internal Census Data." *Rural Sociology* 72, no. 4 (2007): 537–59.

Dun & Bradstreet. *Comprehensive Business Report: Torrey Farms, Inc.*, 2010.

Duncan, James. "Landscape Taste as a Symbol of Group Identity." *Geographical Review* 63, no. 3 (1973): 334–55.

DuPuis, E. Melanie. "Angels and Vegetables: A Brief History of Food Advice in America." *Gastronomica: The Journal of Food and Culture* 7, no. 3 (2007): 33–44.

DuPuis, E. Melanie, and David Goodman. "Should We Go 'Home' to Eat? Toward a Reflexive Politics of Localism." *Journal of Rural Studies* 21, no. 3 (2005): 357–71.

Durand, Jorge, Douglas S. Massey, and Emilio A. Parrado. "The New Era of Mexican Migration to the United States." *Journal of American History* 86, no. 2 (1999): 518–36.

Durand, Jorge, Douglas S. Massey, and René M. Zenteno. "Mexican Immigration to the United States: Continuities and Change." *Latin American Research Review* 36, no. 1 (2001): 107–27.

Dyer Jr., William G. *Cultural Change in Family Firms: Anticipating and Managing Business and Family Transitions*. San Francisco: Jossey-Bass, 1986.

Edid, Maralyn. *Agricultural Labor Markets in New York State and Implications for Labor Policy*. Cornell University Task Force on Farmworker Collective Bargaining, May 1991.

———. *Farm Labor Organizing: Trends and Prospects*. Ithaca, NY: Cornell University Press, 1994.

Eisinger, Chester E. "The Freehold Concept in Eighteenth-Century American Letters." *William and Mary Quarterly* 4, no. 1 (1947): 42–59.

Ellis, David M. "Land Tenure and Tenancy in the Hudson Valley, 1790–1860." *Agricultural History* 18, no. 2 (1944): 75–82.

Estabrook, Barry. *Tomatoland: How Modern Agriculture Destroyed Our Most Alluring Fruit*. Kansas City, MO: Andrews McMeel, 2011.

Farmer, Frank L., and Zola K. Moon. "An Empirical Examination of Characteristics of Mexican Migrants to Metropolitan and Nonmetropolitan Areas of the United States." *Rural Sociology* 74, no. 2 (2009): 220–40.

Feagan, Robert, and Amanda Henderson. "Devon Acres CSA: Local Struggles in a Global Food System." *Agriculture Human Values* 26, no. 3: 203–17.

Feenstra, Gail W. "Local Food Systems and Sustainable Communities." *American Journal of Alternative Agriculture* 12, no. 1 (1997): 28–36.

Ferguson, Kirsten. *Saving Working Landscapes: A Campaign for Hudson Valley Farms*. American Farmland Trust, 2002.

Fisher, Lloyd Horace. *The Harvest Labor Market in California*. Cambridge, MA: Harvard University Press, 1953.

Flinn, William L., and Donald E. Johnson. "Agrarianism among Wisconsin Farmers." *Rural Sociology* 39, no. 2 (1974): 187–204.

Fojo, Carolina, Dayna Burtness, and Vera Chang. *Inventory of Farmworker Issues and Protections in the United States.* Palo Alto, CA: Bon Appétit Management Company Foundation and The United Farmworkers, 2011. www.bamco.com /uploads/documents/farmworkerinventory_0428_2011.pdf.

Foner, Nancy. "What's New about Transnationalism? New York Immigrants Today and at the Turn of the Century." *Diaspora* 6, no. 3 (1999): 355–75.

Foner, Phillip S., and Ronald L. Lewis, eds. *Black Workers: A Documentary History from Colonial Times to the Present.* Philadelphia: Temple University Press, 1989.

Friedland, William H. "Labor Waste in New York: Rural Exploitation and Migrant Workers." *Trans-Action* 6, no. 4 (1969): 48–53.

———. *Manufacturing Green Gold: Capital, Labor and Technology in the Lettuce Industry.* New York: Cambridge University Press, 1981.

Friedland, William H., and Dorothy Nelkin. *Migrant: Agricultural Workers in America's Northeast.* New York: Holt, Rinehart and Winston, 1971.

Friend, Catherine. *Hit by a Farm: How I Learned to Stop Worrying and Love the Barn.* Cambridge, MA: Da Capo Press, 2006.

Galarza, Ernesto. *Merchants of Labor: The Mexican Bracero Story; An Account of the Managed Migration of Mexican Farm Workers in California, 1942–1960.* Santa Barbara, CA: McNally & Loftin, 1964.

Galbraith, John Kenneth. *The Nature of Mass Poverty.* Cambridge, MA: Harvard University Press, 1979.

Ganz, Marshall. *Why David Sometimes Wins: Leadership, Organization, and Strategy in the California Farm Worker Movement.* New York: Oxford University Press, 2009.

Garcia, Matthew. "Labor, Migration, and Social Justice in the Age of the Grape Boycott." *Gastronomica: The Journal of Food and Culture* 7, no. 3 (2007): 68–74.

García-Colón, Ismael. "Claiming Equality: Puerto Rican Farmworkers in Western New York." *Latino Studies* 6, no. 3 (2008): 269–89.

Gassan, Richard H. *The Birth of American Tourism: New York, the Hudson Valley, and American Culture, 1790–1830.* Amherst, MA: University of Massachusetts Press, 2008.

Geffert, Garry G. "H-2A Guestworker Program." In *The Human Cost of Food: Farmworkers' Lives, Labor, and Advocacy,* edited by Charles D. Thompson Jr. and Melinda Wiggins, 113–35. Austin: University of Texas Press, 2002.

Genovese, Eugene D. *Roll, Jordan, Roll: The World the Slaves Made.* New York: Pantheon Books, 1976.

George-Warren, Holly, ed. *Farm Aid: A Song for America.* Emmaus, PA: Rodale Books, 2005.

Gold, Michael Evan. *An Introduction to Labor Law, Revised Edition.* Ithaca, NY: Cornell University Press, 1998.

Goldfarb, Ronald L. *Migrant Farm Workers: A Caste of Despair.* Ames: Iowa State University Press, 1981.

Goodell, Grace E. "Paternalism, Patronage, and Potlatch: The Dynamics of Giving and Being Given To." *Current Anthropology* 26, no. 2 (1985): 247–66.

Grady, Sarah. *The State of Agriculture in the Hudson Valley.* Cold Spring, NY: Glynwood Institute, 2010. www.glynwood.org/publications-multimedia /state-of-ag/.

Gray, Margaret. "How Latin American Inequality Becomes Latino Inequality: A Case Study of Hudson Valley Farmworkers." In *Indelible Inequalities in Latin America: Insights from History, Politics and Culture,* edited by Paul Gootenberg and Luis Reygades, 169–92. Durham, NC: Duke University Press, 2010.

———. *The Hudson Valley Farmworker Report: Understanding the Needs and Aspirations of a Voiceless Population.* Annandale-on-Hudson, NY: Bard College, 2007.

———. "Mechanics of Empowerment: Migrant Farmworker Advocacy." In *Global Migration, Social Change, and Cultural Transformation,* edited by Emory Eliot, Jasmine Payne, and Patricia Ploesch, 207–24. New York: Palgrave-Macmillan, 2007.

Gray, Margaret, and Shareen Hertel. "Immigrant Farmworker Advocacy: The Dynamics of Organizing." *Polity* 41, no. 4 (2009): 409–35.

Green, Nancy L. "Women and Immigrants in the Sweatshop: Categories of Labor Segmentation Revisited." *Society for Comparative Study of Society and History* 38, no. 3 (1996): 411–33.

Greenwood, Michael J. "Research on Internal Migration in the United States: A Survey." *Journal of Economic Literature* 13, no. 2 (1975): 397–433.

Griffith, David. *American Guestworkers: Jamaicans and Mexicans in the U.S. Labor Market.* University Park: Pennsylvania State University Press, 2006.

———. *The Estuary's Gift: An Atlantic Coast Cultural Biography.* University Park: Pennsylvania State University Press, 1999.

Griffith, David, and Edward Kissam. *Working Poor: Farmworkers in the United States.* Philadelphia: Temple University Press, 1995.

Griswold, Alfred Whitney. *Farming and Democracy.* New Haven, CT: Yale University Press, 1952.

Grossardt, Ted. "Harvest(ing) Hoboes: The Production of Labor Organization through the Wheat Harvest." *Agricultural History* 70, no. 2 (1996): 283–301.

Gussow, Joan Dye. *The Feeding Web: Issues in Nutritional Ecology.* Palo Alto, CA: Bull Publishing, 1978.

———. *This Organic Life: Confessions of a Suburban Homesteader.* White River Junction, VT: Chelsea Green, 2002.

Guthman, Julie. *Agrarian Dreams: The Paradox of Organic Farming in California.* Berkeley: University of California Press, 2004.

Habraken, N.J. *The Structure of the Ordinary.* Cambridge, MA: MIT Press, 1998.

Hahamovitch, Cindy. *The Fruits of Their Labor: Atlantic Coast Farmworkers and the Making of Migrant Poverty, 1870–1945.* Chapel Hill: University of North Carolina Press, 1997.

Halweil, Brian. *Eat Here: Reclaiming Homegrown Pleasures in a Global Supermarket.* Washington, DC: Worldwatch Institute, 2004.

Hamilton, Emily Rebecca. "Farmworker Housing in New York State: Obstacles and Options 2010." MA thesis, Cornell University, Department of City and Regional Planning, 2010.

Hand, Michael, and Stephen Martinez. "Just What Does Local Mean?" *Choices Magazine* 25, no. 1 (2010). www.choicesmagazine.org/magazine/article. php?article=108.

Hanson, Victor Davis. *Fields Without Dreams: Defending the Agrarian Idea.* New York: The Free Press, 1996.

———. *The Other Greeks: The Family Farm and the Agrarian Roots of Western Civilization.* Berkeley: University of California Press, 1995.

———. "A Secretary for Farmland Security." *New York Times,* December 9, 2004. www.nytimes.com/2004/12/09/opinion/09hanson.html.

Harrison, Jill. "Lessons Learned from Pesticide Drift: A Call to Bring Production Agriculture, Farm Labor and Social Justice Back into Agrifood Research and Activism." *Agriculture and Human Values* 25, no. 2 (2008): 163–67.

Harvey, David. *Justice, Nature and the Geography of Difference.* Oxford: Blackwell, 1996.

Haslip-Viera, Gabriel, and Sherrie L. Baver. *Latinos in New York: Communities in Transition.* Notre Dame, IN: University of Notre Dame Press, 1996.

Hatton, Timothy J., and Jeffrey G. Williamson. "What Drove the Mass Migrations from Europe in the Late Nineteenth Century?" *Population and Development Review* 20, no. 3 (1994): 533–59.

Hawley, Stephen. *Hawley Stands Opposed to Farm Death Bill.* New York State Assembly, June 8, 2009. http://assembly.state.ny.us/mem/Stephen-Hawley /story/32496/.

Heffernan, William D., and Judith Bortner Heffernan. "Impact of the Farm Crisis on Rural Families and Communities." *Rural Sociologist* 6, no. 3 (1986): 160–70.

Herbert, Bob. "State of Shame." *New York Times,* June 8, 2009. www.nytimes. com/2009/06/09/opinion/09herbert.html.

Hernández-León, Rubén. *Metropolitan Migrants.* Berkeley: University of California Press, 2008.

Hightower, Jim. *Hard Tomatoes, Hard Times: A Report of the Agribusiness Accountability Project on the Failure of America's Land Grant College Complex.* New York: Schenkman, 1973.

Hinrichs, C. Clare. "Embeddedness and Local Food Systems: Notes on Two Types of Direct Agricultural Market." *Journal of Rural Studies* 16, no. 3 (2000): 295-303.

———. "The Practice and Politics of Food System Localization." *Journal of Rural Studies* 19, no. 1 (2003): 33–45.

Hinrichs, C. Clare, Jack Kloppenburg, Steve Stevenson, Sharon Lezberg, John Hendrickson, and Kathy DeMaster. *Moving Beyond Global and Local.* United States Department of Agriculture, Regional Research Project NE-185 working statement, October 2, 1998.

Hofstadter, Richard. *The Age of Reform: From Bryan to F.D.R.* New York: Vintage Books, 1955.

Holbrook, Morris B., and Kim P. Corfman. "Quality and Value in the Consumption Experience: Phaedrus Rides Again." In *Perceived Quality,* edited by Jacob Jacoby and Jerry C. Olson, 31–57. Lexington, MA: Lexington Books, 1985.

Hoppe, Robert A., Penni Korb, Erik J. O'Donoghue, and David E. Banker. "Structure and Finances of U.S. Farms: Family Farm Report, 2007 Edition." *Economic Information Bulletin No. (EIB-24).* U.S. Department of Agriculture, Economic Research Service, 2007.

Horne, Field. "Life on a Rocky Farm." *Hudson Valley Regional Review* 7, no. 1 (1990): 31–41.

Humphrey, Thomas J. *Land and Liberty: Hudson Valley Riots in the Age of Revolution.* DeKalb: Northern Illinois Press, 2004.

Hunt, Alan. "Consumer Interactions and Influences on Farmers' Market Vendors." *Renewable Agriculture and Food Systems* 22, no. 1 (2007): 54–66.

Hunter, Albert. "The Symbolic Ecology of Suburbia." In *Neighborhood and Community Environments,* edited by Irwin Altman and Abraham Wandersman, 191–221. New York: Plenum Press, 1987.

Hurd, T. N. *New York's Harvest Labor.* Interdepartmental Committee on Farm and Food Processing Labor, 1953.

Izcara Palacios, Simón Pedro. "La adicción a la mano de obra ilegal: Jornaleros tamaulipecos en Estados Unidos." *Latin American Research Review* 45, no. 1 (2010): 55–75.

Jackman, Mary. *The Velvet Glove: Paternalism and Conflict in Gender, Class, and Race Relations.* Berkeley: University of California Press, 1994.

Jacques, Sean, and Lyn Collins. "Community Supported Agriculture: An Alternative to Agribusiness." *Geography Review* 16, no. 5 (2003): 30–33.

Jayaraman, Saru. *The Hands That Feed Us: Challenges and Opportunities for Workers along the Food Chain.* Food Chain Workers Alliance, June 6, 2012.

Jefferson, Thomas. *Notes on the State of Virginia.* Richmond, VA: J. W. Randolph, 1853.

Jensen, Leif. *New Immigrant Settlements in Rural America: Problems, Prospects, and Policies.* Carsey Institute, University of New Hampshire, 2006.

Johnson, Kirk. "Farmers Strain to Hire American Workers in Place of Migrant Labor." *New York Times,* October 5, 2011. www.nytimes.com/2011/10/05/us/farmers-strain-to-hire-american-workers-in-place-of-migrant-labor.html.

Johnson, Liz. "Garden of Eating." *Arrive: The Magazine for Northeast Business Travelers,* October 2008.

Johnson, Tricia. "Looking for Your Own Dream Farm?" *LIFE,* September 29, 2006.

Johnston, Anne. "Farm Worker Killed on Job." *Daily Messenger,* November 10, 2004.

Johnstone, Paul H. "Old Ideals versus New Ideas in Farm Life." In *1940 Yearbook of Agriculture: Farmers in a Changing World,* 111–70. Washington, DC: United States Department of Agriculture, United States Printing Office, 1940.

Kabeer, Naila. *The Power to Choose: Bangladeshi Women and Labor Market Decisions in London and Dhaka*. New York: Verso, 2000.

Kandel, William, and John Cromartie. *New Patters of Hispanic Settlement in Rural America*. Washington, DC: U.S. Department of Agriculture, Economic Research Service, 2004.

Kelley, Robin D. G. *Race Rebels: Culture, Politics, and the Black Working Class*. New York: The Free Press, 1994.

Kennedy, Kerry. "This Must Be New York's Final Harvest of Shame: Let Us Finally Protect Exploited Farmworkers." *New York Daily News*, November 26, 2009. www.nydailynews.com/opinions/2009/11/26/2009-11-26_this_must_be_new_yorks_final_harvest_of_shame.html.

Kilmer-Purcell, Josh. *The Bucolic Plague: How Two Manhattanites Became Gentlemen Farmers: An Unconventional Memoir*. New York: HarperCollins Publishers, 2010.

Kimball, Kristin. *The Dirty Life: A Memoir of Farming, Food, and Love*. New York: Scribner, 2010.

Kimbrell, Andrew. *The Fatal Harvest Reader: The Tragedy of Industrial Agriculture*. Washington, DC: Island Press, 2002.

Kingsolver, Barbara. *Animal, Vegetable, Miracle: A Year of Food Life*. New York: HarperCollins, 2007.

Klinkenborg, Verlyn. *Making Hay*. Guilford, CT: Lyons Press, 1997.

Knights, David, and Darren McCabe. "'A Different World': Shifting Masculinities in the Transition to Call Centres." *Organization* 8, no. 4 (2001): 619–45.

Kochhar, Rakesh, Roberto Suro, and Sonya Tafoya. *The New Latino South: The Context and Consequences of Rapid Population Growth*. Washington, DC: Pew Hispanic Center, 2005.

Krissam, Fred. "Sin Coyote Ni Patrón: Why the 'Migrant Network' Fails to Explain International Migration." *International Migration Review* 39, no. 1 (2005): 4–43.

Kristof, Nicholas D. "Food for the Soul." *New York Times*. August 22, 2009. www.nytimes.com/2009/08/23/opinion/23kristof.html.

Kummer, Corby. "The Great Grocery Smackdown." *Atlantic*, March 2010.

Lapp, Michael. "Managing Migration: The Migration Division of Puerto Rico and Puerto Ricans in New York City, 1948–1968." PhD diss., Johns Hopkins University, 1991.

Lappé, Anna, and Bryant Terry. *Grub: Ideas for an Urban Organic Kitchen*. New York: Penguin, 2006.

Lappé, Frances Moore. *Diet for a Small Planet*. New York: Ballantine Books, 1990.

Lappé, Frances Moore, and Anna Lappé. *Hope's Edge: The Next Diet for a Small Planet*. New York: Jeremy P. Tarcher, 2003.

Larson, Olaf. *Migratory Agricultural Workers in the Eastern Seaboard States, Rural Poverty in the United States*. President's National Advisory Committee on Rural Poverty. Government Printing Office, 1968.

Lauman, Edward, and James House. "Living Room Styles and Social Attributes: The Patterning of Material Artifacts in a Modern Urban Community." *Sociology and Social Research* 54, no. 3 (1970): 321–42.

Lehmann, Joyce Woelfle. *Migrant Farmworkers of Wayne County, New York: A Collection of Oral Histories from the Back Roads.* Lyons, NY: Wayne County Historical Society, 1990.

Leiter, Jeffrey, Michael D. Schulman, and Rhonda Zingraft, eds. *Hanging by a Thread: Social Change in Southern Textiles.* Ithaca, NY: Cornell University Press, 1991.

Leopold, Aldo. *Sand County Almanac.* New York: Oxford University Press, 1949.

Lev, Larry, and Lauren Gwin. "Filling in the Gaps: Eight Things to Recognize about Farm-Direct Marketing." *Choices Magazine* 25, no. 1 (2010). www.choicesmagazine. org/magazine/article.php?article=110.

Lewis, R. W. B. *The American Adam: Innocence, Tragedy, and Tradition in the Nineteenth Century.* Chicago: University of Chicago Press, 1955.

Lewthwaite, Stephanie. "Race, Paternalism, and 'California Pastoral': Rural Rehabilitation and Mexican Labor in Greater Los Angeles." *Agricultural History* 81, no. 1 (2007): 1–35.

Lichter, Daniel T. "Immigration and the New Racial Diversity in Rural America." *Rural Sociology* 77, no. 1 (2012): 3–35.

Lichter, Daniel T., and Kenneth M. Johnson. "Emerging Rural Settlement Patterns and the Geographic Redistribution of America's New Immigrants." *Rural Sociology* 1, no. 1 (2006): 109–31.

Linder, Marc. *Migrant Workers and Minimum Wages: Regulating the Exploitation of Agricultural Labor in the United States.* Boulder: Westview Press, 1992.

Low, Sarah A., and Stephen Vogel. *Direct and Intermediate Marketing of Local Foods in the United States.* ERR-128. Washington, DC: U.S. Department of Agriculture, Economic Research Service, November 2011.

Lyson, Thomas A. *Civic Agriculture: Reconnecting Farm, Food, and Community.* Medford, MA: Tufts University Press, 2004.

Lyson, Thomas A., and Judy Green. "The Agricultural Marketscape: A Framework for Sustaining Agriculture and Communities in the Northeast." *Journal of Sustainable Agriculture* 15, no. 2–3 (1999): 133–50.

Magdoff, Fred, John Bellamy Foster, and Frederick H. Buttel. *Hungry for Profit: The Agribusiness Threat to Farmers, Food, and the Environment.* New York: New York University Press, 2000.

Mahler, Sarah J. *American Dreaming: Immigrant Life on the Margins.* Princeton, NJ: Princeton University Press, 1995.

Majka, Linda C., and Theodore J. Majka. *Farm Workers, Agribusiness, and the State.* Philadelphia: Temple University Press, 1982.

———. "Organizing U.S. Farm Workers: A Continuous Struggle." In *Hungry for Profit: The Agribusiness Threat to Farmers, Food, and the Environment,* edited by Fred Magdoff, John Bellamy Foster, and Frederick H. Buttel, 161–74. New York: Monthly Review Press, 2000.

Maldonado, Marta Maria. "Racial Triangulation of Latino/a Workers by Agricultural Employers." *Human Organization* 65, no. 4 (2006): 353–61.

Maloney, Thomas R. "Understanding the Dimensions of Culture: Learning to Relate to Hispanic Employees." In *Managing the Hispanic Workforce 2005.* State College, PA: Penn State College of Agricultural Sciences, 2005.

Maloney, Thomas R. and David C. Grusenmeyer. *Survey of Hispanic Dairy Workers in New York State.* Ithaca, NY: Cornell University, February 2005.

Maloney, Thomas R., and Nelson L. Bills. *The New York State Agricultural Immigration and Human Resource Management Issues Study.* Ithaca, NY: Department of Applied Economics and Management, College of Agriculture and Life Sciences, Cornell University, August 2008.

Margolies, Ken. *Training Needs Assessment of Farm Workers in Orange and Sullivan Counties, NY.* Ithaca, NY: Cornell University ILR School, June 20, 2001.

Markus, Thomas A. *Buildings and Power: Freedom and Control in the Origin of Modern Building Types.* London: Routledge, 1993.

Martin, Andrew. "Is a Food Revolution Now in Season?" *New York Times,* March 21, 2009. www.nytimes.com/2009/03/22/business/22food.html?pagewanted=all.

Martin, Philip. "Calculating the Costs and Benefits." *New York Times,* September 30, 2011. www.nytimes.com/roomfordebate/2011/08/17/could-farms-survive-without-illegal-labor/the%20-costs-and-benefits-of-a-raise-for-field-workers.

———. "The Endless Debate: Immigration and US Agriculture." In *The Debate in the United States over Immigration,* edited by Peter Duigan and Lewis Gann, 79–101. Stanford, CA: Hoover Institution, 1998.

———. *Illegal Immigration and the Colonization of the American Labor Market.* Washington, DC: Center for Immigration Studies, January 1986. www.cis.org /AmericanLaborMarket%2526Immigration.

———. "Mexican Workers and U.S. Agriculture: The Revolving Door." *International Migration Review* 36, no. 4 (2002): 1124–42.

Martinez, Steve, Michael Hand, Michelle DaPra, Susan Pollack, Katherine Ralston, Travis Smith, Stephen Vogel, et al. *Local Food Systems: Concepts, Impacts, and Issues.* U.S. Department of Agriculture, Economic Research Service, May 2010.

Massey, Douglas S. "International Migration and Economic Development in Comparative Perspective." *Population and Development Review* 14, no. 3 (1988): 383–414.

———. "The New Geography of Mexican Immigration." In *New Destinations of Mexican Immigration in the United States: Community Formation, Local Responses and Inter-group Relations,* edited by Victor Zúñiga and RubénHernández-León, 1–20. New York: Russell Sage Foundation, 2005.

Massey, Douglas S., Rafael Alarcon, Jorge Durand, and Humberto González. *Return to Aztlan: The Social Process of International Migration from Western Mexico.* Berkeley: University of California Press, 1987.

Massey, Douglas S., and Nancy A. Denton. *American Apartheid: Segregation and the Making of the Underclass.* Cambridge, MA: Harvard University Press, 1993.

Matteson, Jay. "Interview with Julie Suarez." Jefferson County Agricultural Development Corporation and 790 WTNY (Stephens Media Group, Watertown LLC), May 16, 2009.

Matthews, Kathryn. "In Hudson Valley, Farmers' Markets Offer Local Food and Local Chat." *New York Times,* August 14, 2009. http://travel.nytimes.com/2009/08/14/travel/14Hudson.html.

Mazuzan, George T., and Nancy Walker. "Restricted Areas: German Prisoner-of-War Camps in Western New York, 1944–1946." *New York History* 59, no. 1 (1978): 55–72.

McConnell, Grant. *The Decline of Agrarian Democracy.* Berkeley: University of California Press, 1953.

McCorkindale, Susan. *Confessions of a Counterfeit Farm Girl.* New York: New American Library, 2008.

McCurdy, Charles W. *The Anti-Rent Era in New York Law and Politics, 1839–1865.* Chapel Hill: University of North Carolina Press, 2001.

McDermott, William P. *Dutchess County's Plain Folks: Enduring Uncertainty, Inequality, and Uneven Prosperity, 1725–1875.* Clinton Corners, NY: Kerleen Press, 2004.

McKibben, Bill. *Deep Economy: The Wealth of Communities and the Durable Future.* New York: Henry Holt, 2007.

McWilliams, Carey. *Factories in the Field: The Story of Migratory Farm Labor in California.* Boston: Little, Brown and Company, 1935.

Meister, Dick, and Anne Loftis. *A Long Time Coming: The Struggle to Unionize America's Farm Workers.* New York: Macmillan Publishing, 1977.

Mendez, Olga, and Hector Diaz. *Separate & Unequal: New York's Farmworkers.* New York State Senate-Assembly Puerto Rican / Hispanic Task Force, Joint Temporary Task Force on Farmworker Issues, April 1995.

Merrill, Richard, ed. *Radical Agriculture.* New York: Harper Colophon Books, 1976.

Millard, Ann V., and Jorge Chapa. *Apple Pie and Enchiladas: Latino Newcomers in the Rural Midwest.* Austin: University of Texas Press, 2004.

Millones, Peter. "Kennedy and Javits Are Shocked by Housing of Migrants Upstate; Ordered to Leave Kennedy Incredulous." *New York Times,* September 9, 1967.

Mirengoff, William. *Puerto Rican Farm Workers in the Middle Atlantic States: Highlights of a Study.* U.S. Department of Labor, Bureau of Employment Security, Division of Reports and Analysis, November 1954.

Mishra, Ashok K., Hisham S. El-Osta, Mitchell J. Morehart, James D. Johnson, and Jeffrey W. Hopkins. *Income, Wealth, and Economic Well-Being of Farm Households, Agricultural Economic Report No. 812.* Washington, DC: Farm Sector Performance and Well-Being Branch, Resource Economics Division, Economic Research Service, U.S. Department of Agriculture, 2002. http://ageconsearch.umn.edu/bitstream/33967/1/ae020812.pdf.

Mitchell, Don. *The Lie of the Land: Migrant Workers and the California Landscape.* Minneapolis, MN: University of Minnesota Press, 1996.

Molnar, Joseph J., and Litchi S. Wu. "Agrarianism, Family Farming, and Support for State Intervention in Agriculture." *Rural Sociology* 54, no. 2 (1989): 227–45.

Moore, Truman. *The Slaves We Rent.* New York: Random House, 1965.

Morales, Miguel. "How to Make the Most of My Multicultural Workforce." In *42nd Florida Dairy Production Conference.* Gainesville, 2005.

Morin, Alexander. *The Organizability of Farm Labor in the United States.* Cambridge, MA: Harvard University Press, 1952.

Moser, Riccarda, Roberta Raefelli, and Dawn Thilmany. "Identifying Influential Attributes in WTP for Agro-Food with Credence Based Public Good Attributes: The State of the Art and Some Implications for CE." Presented at the Workshop on Valuation Methods in Agro-Food and Environmental Economics, Barcelona, Spain, July 2008.

Moss, Philip, and Chris Tilly. *Stories Employers Tell: Race, Skill, and Hiring in America.* New York: Russell Sage Foundation, 2001.

Mountz, Alison, and Richard A. Wright. "Daily Life in the Transnational Migrant Community of San Agustin, Oaxaca, and Poughkeepsie, New York." *Diaspora* 5, no. 3 (1996).

Murdoch, Jonathan, Terry Marsden, and Jo Banks. "Quality, Nature, and Embeddedness: Some Theoretical Considerations in the Context of the Food Sector." *Economic Geography* 76, no. 2 (2000): 107–25.

Murphy, Arthur D., Colleen Blanchard, and Jennifer A. Hill. *Latino Workers in the Contemporary South.* Athens, GA: University of Georgia Press, 2001.

Nabhan, Gary Paul. *Coming Home to Eat: The Pleasures and Politics of Local Foods.* New York: W. W. Norton Press, 2002.

Nelkin, Dorothy. *On the Season: Aspects of the Migrant Labor System.* Ithaca, NY: New York State School of Industrial and Labor Relations, Cornell University, 1970.

———. "Response to Marginality: The Case of Migrant Farm Workers." *British Journal of Sociology* 20, no. 4 (1969): 375–89.

Nestle, Marion. *Food Politics: How the Food Industry Influences Nutrition and Health.* Berkeley: University of California Press, 2003.

———. *What to Eat.* New York: North Point Press, 2006.

Nestle, Marion, and W. Alex McIntosh. "Writing the Food Studies Movement." *Food, Culture & Society* 13, no. 2 (2010): 159–68.

Neuman, Kristin E., and Marta Tienda. "The Settlement and Secondary Migration Patterns of Legalized Immigrants: Insights from Administrative Records." In *Immigration and Ethnicity: The Integration of America's Newest Immigrants,* edited by Barry Edmonston and Jeffrey S. Passel, 187–226. Lanham, MA: Urban Institute, 1994.

Neuman, William, and David Barboza. "U.S. Drops Inspector of Food in China." *New York Times,* June 13, 2010. www.nytimes.com/2010/06/14/business /global/14organic.html.

Nevaer, Louis, and Vaso Perimenis Ekstein. *HR and the New Hispanic Workforce: A Comprehensive Guide to Cultivating and Leveraging Employee Success.* Ithaca, NY: Cornell Cooperative Extension, 2009.

New York Farm Bureau. *Farmworker Bill Defeated.* Albany: New York Farm Bureau, August 4, 2010. archive.is/D5sD.

————. *Letter Regarding Farmworkers March from Auburn to Albany.* Albany, NY: New York Farm Bureau, 2004. www.nyfb.org/features/march.htm.

Newby, Howard. "The Deferential Dialectic." *Comparative Studies in Society and History* 17, no. 2 (1975): 139–64.

Newby, Howard, Colin Bell, David Rose, and Peter Saunders. *Property, Paternalism and Power: Class and Control in Rural England.* London: Hutchinson of London, 1978.

Nieves, Evelyn. "Farm Workers Stand Tough and Walk Out." *New York Times,* November 3, 1996. www.nytimes.com/1996/11/03/nyregion/farm-workers-stand-tough-and-walk-out.html

Niles, Daniel, and Robin Jane Roff. "Shifting Agrifood Systems: The Contemporary Geography of Food and Agriculture; an Introduction." *GeoJournal* 71, no. 1 (2008): 1-10.

Norberg-Hodge, Helena, Todd Merrifield, and Steven Gorelick. *Bringing the Food Economy Home: Local Alternatives to Global Agribusiness.* London: Zed, 2002.

Norden, Larry, David E. Pozen, and Bethany L. Foster. *Unfinished Business: New York Legislative Reform.* Brennan Center for Justice, 2006.

Norris, G. M. "Industrial Paternalist Capitalism and Local Labour Markets." *Sociology* 12, no. 3 (1978): 469–89.

Norris, Linda, and Elaine D. Engst. *Migrant Farmworkers Records in Upstate New York: Survey and Guide.* Ithaca, NY: Cornell University Library, 1999.

North, David, and J. S. Holt. "The Apple Industry in New York and Pennsylvania." In Appendix 1: *Case Studies and Research Reports Prepared for the Commission on Agricultural Workers 1989–1993,* 369–443. Washington, DC: U.S. Government Printing Office, 1993.

Norton, Dean. "Don't Turn Farms into Factories." *Grassroots,* February 2010, 1, 23. www.nyfb.org/img/uploads/file/Grassroots-Feb10-forweb.pdf.

Novesky, Jerry, and Janet Crawshaw. "Branding the Region." *Valley Table,* November 2009. http://valleytable.com/article.php?article=004+Features%2F Branding+the+region.

Oberholtzer, Lydia, and Shelly Grow. *Producer-only Farmers' Markets in the Mid-Atlantic Region: A Survey of Market Managers.* Arlington, VA: Henry A. Wallace Center for Agricultural and Environmental Policy at Winrock International, 2003.

Oliver, Melvin L., and Thomas M. Shapiro. *Black Wealth/White Wealth: A New Perspective on Racial Inequality.* New York: Routledge, 1995.

Ostrom, Marcia. "Everyday Meanings of 'Local Food': Views from Home and Field." *Community Development* 37, no. 1 (2006): 65–78.

Oxfam America. *Like Machines in the Fields: Workers without Rights in American Agriculture.* Oxfam America, March 2004.

Paarlberg, Don. *Farm and Food Policy: Issues of the 1980s.* Lincoln: University of Nebraska Press, 1980.

Pacione, Michael. "Local Exchange Trading Systems—A Rural Response to the Globalization of Capitalism?" *Journal of Rural Studies* 13, no. 4 (1997): 415–27.

Padavic, Irene, and William R. Earnest. "Paternalism as a Component of Managerial Strategy." *Social Science Journal* 31, no. 4 (1994): 389–405.

Pager, Devah. "The Mark of a Criminal Record." *American Journal of Sociology* 108, no. 5 (2003): 937–75.

Papademetriou, Demetrios G., and Gerald W. Hopple. "Causal Modeling in International Migration Research: A Methodological Prolegomenon." *Quality and Quantity* 16, no. 5 (1982): 369–402.

Parkerson, Donald H. *The Agricultural Transition in New York State: Markets and Migration in Mid-Nineteenth-Century America*. Ames: Iowa State University Press, 1995.

Patel, Raj. *Stuffed and Starved: Markets, Power and the Hidden Battle for the World Food System*. London: Portobello Books, 2007.

Peck, Jamie. *Work Place: The Social Regulation of Labor Markets*. New York: Guilford Press, 1996.

Pedersen, Donald B. "Introduction to the Agricultural Law Symposium." *University of California, Davis Law Review* 23, no. 3 (1990): 401–14.

Petrini, Carlo. *Slow Food (The Case for Taste)*. New York: Columbia University Press, 2003.

Pfeffer, Max J., and Pilar A. Parra. *Immigrants and the Community*. Ithaca, NY: Development Sociology Department, College of Agriculture and Life Sciences, Cornell University, 2004.

———. *Immigrants and the Community: Community Perspectives*. Ithaca, NY: Development Sociology Department, College of Agriculture and Life Sciences, Cornell University, 2005.

———. *Immigrants and the Community: Farmworkers with Families*. Ithaca, NY: Development Sociology Department, College of Agriculture and Life Sciences, Cornell University, 2005.

———. *Immigrants and the Community: Former Farmworkers*. Ithaca, NY: Development Sociology Department, College of Agriculture and Life Sciences, Cornell University, 2005.

———. "New Immigrants in Rural Communities: The Challenges of Integration." *Social Text* 24, no. 3 (2006): 81–98.

———. "Strong Ties, Weak Ties and Human Capital: Latino Immigrant Employment Outside the Enclave." *Rural Sociology* 74, no. 2 (2009): 241–69.

Phillips, Julie A., and Douglas S. Massey. "The New Labor Market: Immigrants and Wages after IRCA." *Demography* 36, no. 2 (1999): 233–46.

Piore, Michael J. *Birds of Passage: Migrant Labor and Industrial Societies*. London: Cambridge University Press, 1979.

Pirog, Rich, Timothy Van Pelt, Kamyar Enshayan, and Ellen Cook. *Food, Fuel, and Freeways: An Iowa Perspective on How Far Food Travels, Fuel Usage, and Greenhouse Gas Emissions*. Leopold Center for Sustainable Agriculture, Iowa State University, June 2001.

Polikoff, Alexander. "Racial Inequality and the Black Ghetto." *Poverty & Race* 13, no. 6 (2004): 1–10.

Pollan, Michael. *The Botany of Desire.* New York: Random House, 2001.

———. *The Omnivore's Dilemma: A Natural History of Four Meals.* New York: Penguin Press, 2006.

———. "Vote for the Dinner Party." *New York Times Magazine,* October 10, 2012. www.nytimes.com/2012/10/14/magazine/why-californias-proposition-37-should-matter-to-anyone-who-cares-about-food.html?WT.mc_id=NYT-L-P-FOOD-MAG-101412-L4.

Portes, Alejandro, and Robert L. Bach. *Latin Journey: Cuban and Mexican Immigrants in the United States.* Berkeley: University of California Press, 1985.

Portes, Alejandro, and John Watson. *Labor, Class, and the International System.* New York: Academic Press, 1981.

Powell, Benjamin. "The Law of Unintended Consequences: Georgia's Immigration Law Backfires." *Forbes,* May 17, 2012. www.forbes.com/sites/realspin/2012/05/17/the-law-of-unintended-consequences-georgias-immigration-law-backfires/.

Pratt, Gerry. "The House as an Expression of Social Worlds." In *Housing and Identity: Cross-Cultural Perspectives,* edited by James Duncan, 135–80. New York: Holmes and Meier, 1982.

Pringle, Peter. *Food Inc.: Mendel to Monsanto—The Promises and Perils of the Biotech Harvest.* New York: Simon and Schuster, 2003.

Rehak, Melanie. "Growing Pains." *Bookforum,* Summer 2010.

Reich, Michael, David M. Gordon, and Richard C. Edwards. "A Theory of Labor Market Segmentation." *American Economic Review* 63, no. 2 (1973): 359–65.

Review Committee. *Internal Review of the Cornell Migrant Program: A Report to the Deans of the Colleges of Human Ecology and Agriculture and Life Sciences and the Director of Cornell Cooperative Extension.* Cornell University, June 2, 2003.

Rhyne, Jennings J. *Some Southern Cotton Mill Workers and Their Villages.* Chapel Hill: University of North Carolina Press, 1930.

Rifkin, Jeremy. *Beyond Beef.* New York: Plume, 1992.

Rivera, Alex. *The Sixth Section.* Directed by Alex Rivera. Brooklyn, NY: American Documentary, P.O.V., 2004.

Rivera-Batiz, Francisco. "Undocumented Workers in the Labor Market: An Analysis of Earnings of Legal and Illegal Mexican Immigrants in the United States." *Journal of Population Economics* 12, no. 1 (1999): 91–116.

Rivers, Tom. *Farm Hands: Hard Work and Hard Lessons from Western New York Fields.* Batavia, NY: Hodgins Printing, 2010.

Rohrer, Wayne C. "Agrarianism and the Social Organization of U.S. Agriculture: The Concomitance of Stability and Change." *Rural Sociology* 35, no. 1 (1970): 5–14.

Rohrer, Wayne C., and Louis H. Douglas. *The Agrarian Transition in America: Dualism and Change.* New York: Bobbs-Merrill, 1969.

Rothenberg, Daniel. *With These Hands: The Hidden World of Migrant Farmworkers Today.* Berkeley: University of California Press, 2000.

Russell, Steven. "Small Farm Dreams." *Life,* September 29, 2006.

Rytina, Nancy. *IRCA Legalization Effects: Lawful Permanent Residence and Naturalization through 2001*. Office of Policy and Planning, Statistics Division, U.S. Immigration and Naturalization Service, October 25, 2002.

Sánchez, Guadalupe L., and Jesús Romo. *Organizing Mexican Undocumented Farm Workers on Both Sides of the Border*. San Diego: Program in United States–Mexican Studies, University of California at San Diego, 1981.

Sander, Libby. "Source of Deadly E. Coli Is Found." *New York Times,* October 13, 2006. www.nytimes.com/2006/10/13/us/13spinach.html.

Sassen, Saskia. *Guests and Aliens*. New York: The New Press, 1999.

———. *The Mobility of Labor and Capital: A Study in International Investment and Labor Flow*. Cambridge: Cambridge University Press, 1988.

Schauer, Robert F., and Dennis G. Tyler. "The Unionization of Farm Labor." *University of California, Davis Law Review* 2, no. 1 (1970): 1–38.

Schlosser, Eric. *Fast Food Nation*. Boston: Houghton Mifflin, 2001.

———. "Penny Foolish." *New York Times,* November 29, 2007. www.nytimes.com/2007/11/29/opinion/29schlosser.html

———. *Reefer Madness: Sex, Drugs, and Cheap Labor in the American Black Market*. Boston: Mariner Books, 2003.

Schnell, Steven M. "Food with a Farmer's Face: Community-Supported Agriculture in the United States." *Geographical Review* 97, no. 4 (2007): 550–64.

Schreck, Aimee, Christy Getz, and Gail Feenstra. "Farmworkers in Organic Agriculture: Toward a Broader Notion of Sustainability." *Sustainable Agriculture* 17, no. 1 (2005). www.sarep.ucdavis.edu/newsltr/v17n1/sa-1.htm.

Schumacher, E. F. *Small Is Beautiful: Economics as if People Mattered*. New York: Harper & Row, 1973.

Scranton, Philip. "Varieties of Paternalism: Industrial Structures and the Social Relations of Production in American Textiles." *American Quarterly* 36, no. 2 (1984): 235–57.

Semple, Kirk. "A Killing in a Town Where Latinos Sense Hate." *New York Times,* November 14, 2008. www.nytimes.com/2008/11/14/nyregion/14immigrant.html?pagewanted=all

Severson, Kim. "Eating: Why Roots Matter More." *New York Times,* November 15, 2006. www.nytimes.com/2006/11/15/business/smallbusiness/15recall.html?pagewanted=all.

Sheingate, Adam D. "Institutions and Interest Group Power: Agricultural Policy in the United States, France, and Japan." *Studies in American Political Development* 14, no. 2 (2000): 184–211.

Sherwood, Julie. "The New Migrants." *Daily Messenger,* April 19, 2003.

Shiva, Vandana. *Stolen Harvest: The Hijacking of the Global Food Supply*. Cambridge, MA: South End Press, 2000.

Shteyngart, Gary. "Escape to New York's Hudson Valley." *Travel + Leisure,* October 2010. www.travelandleisure.com/articles/escape-to-new-yorks-hudson-valley/1.

Sinclair, Upton. *The Jungle*. Chicago: Doubleday, Page & Co, 1906.

———. "What Life Means to Me." *Cosmopolitan Magazine,* October 1906.

Singer, Edward Gerald, and Ivan Sergio Freire de Sousa. "The Sociopolitical Consequences of Agrarianism Reconsidered." *Rural Sociology* 48, no. 2 (1983):
291–307.

Singer, Peter. *The Ethics of What We Eat: Why Our Food Choices Matter.* Emmaus,
PA: Rodale Press, 2007.

Sjaastad, L. A. "The Costs and Returns of Human Migration." *Journal of Political
Economy* 70, no. 5 (1962): 80–93.

Smith, Robert C. "Mexicans in New York: Membership and Incorporation in an
Immigrant Community." In *Latinos in New York: Communities in Transition,*
edited by Gabriel Haslip-Viera and Sherrie L. Baver, 57–103. Notre Dame, IN:
University of Notre Dame Press, 1996.

Sorenson, A. Ann, Richard P. Green, and Karen Russ. *Farming on the Edge.* DeKalb,
IL: American Farmland Trust and Center for Agriculture in the Environment,
Northern Illinois University, 1997.

Stanton, B. F. *The Changing Landscape of New York Agriculture in the Twentieth
Century.* Ithaca, NY: Department of Agricultural Economics, New York State
College of Agriculture and Life Sciences, Cornell University, March 1992.

Steinberg, Stephen. "Immigration, African Americans, and Race Discourse." *New
Politics* 10, no. 3 (2005): 42–54.

Stengel, Andrew, Lawrence Norden, and Laura Seago. *Still Broken: New York State
Legislative Reform 2008 Update.* Brennan Center for Justice, 2009.

Stewart, Keith. *It's a Long Road to a Tomato: Tales of an Organic Farmer Who Quit
the Big City for the (Not So) Simple Life.* New York: Marlowe & Company, 2006.

Stigler, George. "The Theory of Economic Regulation." *Bell Journal of Economics
and Management* 2, no. 1 (1971): 3–21.

Striffler, Steve. *Chicken: The Dangerous Transformation of America's Favorite Food.*
New Haven, CT: Yale University Press, 2005.

Suarez, Julie. "Memo Regarding Farmworker Advocacy Day." Albany, NY: New
York Farm Bureau, April 25, 2003.

Sullivan, Amy. "Columnist Nicholas Kristof." *Time,* February 17, 2010. www.time.
com/time/arts/article/0,8599,1964474,00.html.

Telega, Stanley W., and Thomas R. Maloney. *Legislative Actions on Overtime Pay
and Collective Bargaining and Their Implications for Farm Employers in New York
State, 2009-2010.* Ithaca, NY: Cornell University, Department of Applied Economics and Management, 2010. dyson.cornell.edu/outreach/extensionpdf/2010
/Cornell-Dyson-eb1019.pdf.

———. "Provisions and Politics of the Farmworker Fair Labor Practices Act." In
Northeast Dairy Producers Association (NEDPA) 2010 Proceedings, 141–48. Ithaca,
New York: Cornell University, Pro-Dairy Program, 2010. www.ansci.cornell.
edu/prodairy/nedpa/proceedings/2010/nedpa2010.16.Telega.pdf.

Thomas-Lycklama à Nijeholt, Geertje. *On the Road for Work: Migratory Workers on
the East Coast of the United States.* Boston: Martinus Nijhoff Publishing, 1980.

Thompson, Paul. "Agrarianism and the American Philosophical Tradition."
Agriculture and Human Values 7, no. 1 (1990): 3–8.

Tilly, Charles. *Durable Inequality.* Berkeley: University of California Press, 1998.

Todaro, Michael P. *Internal Migration in Developing Countries: A Review of Theory, Evidence, Methodology and Research Priorities.* Geneva: International Labor Office, 1976.

Tomasek, Robert D. "The Migrant Problem and Pressure Group Politics." *Journal of Politics* 23, no. 2 (1961): 295–319.

Toole, Robert M. "The Role of Agriculture at Hudson Valley Historical Sites." *Hudson Valley Regional Review* 17, no. 2 (2000): 1–15.

Trubek, Amy B. *The Taste of Place: A Cultural Journey into Terroir.* Berkeley: University of California Press, 2008.

Tuttle, Will. *The World Peace Diet: Eating for Spiritual Health and Social Harmony.* New York: Lantern Press, 2005.

United States President's Commission on Migratory Labor. *Migratory Labor in American Agriculture.* Washington, DC: United States Government Printing Office, 1951.

United States Department of Agriculture. *2007 Census of Agriculture.* Washington DC: United States Department of Agriculture, March 13, 2007. www.agcensus.usda.gov.

United States Department of Agriculture, National Agricultural Statistics Service. *Table 1. Historical Highlights: 2007 and Earlier Census Years (New York).* Washington, DC: United States Department of Agriculture, 2007.

United States Department of Health and Human Services. "The 2002 HHS Poverty Guidelines. Washington DC: United States Department of Health and Human Services, November 10, 2002. www.aspe.hhs.gov/poverty/02poverty.htm.

United States Department of Labor. *Migrant Farmworkers: Pursuing Security in an Unstable Labor Market.* Washington DC: United States Department of Labor, Office of the Assistant Secretary for Policy, Office of Program Economics, 1994.

Waldinger, Roger D., and Michael I. Lichter. *How the Other Half Works: Immigration and the Social Organization of Labor.* Berkeley: University of California Press, 2003.

"Wal-Mart Could Get Wounded in Grocery Wars." *Forbes,* January 21, 2011. www.forbes.com/sites/greatspeculations/2011/01/21/wal-mart-could-get-wounded-in-grocery-wars/.

Walsh, Bryan. "America's Food Crisis and How to Fix It." *Time,* August 31, 2009.

Waters, Mary C., and Tomás R. Jiménez. "Assessing Immigrant Assimilation: New Empirical and Theoretical Challenges." *Annual Review of Sociology* 31, no. 1 (2005): 105–25.

Ways, Max. *The Negro and the City.* New York: Time-Life Books, 1968.

Weatherell, Charlotte, Angela Tregear, and Johanne Allinson. "In Search of the Concerned Consumer: UK Public Perceptions of Food, Farming and Buying Local." *Journal of Rural Studies* 19, no. 2 (2003): 233–44.

Westneat, Danny. "The Fruits of Our Labor Absurdity." *Seattle Times,* May 25, 2010.

Williams, W. T. B. *The Negro Exodus from the South.* Washington DC: Division of Negro Economics, U.S. Department of Labor, Government Printing Office, 1919.

Wilson, William J. *When Work Disappears: The World of the New Urban Poor.* New York: Knopf, 1996.

Wingerd, Mary Lether. "Rethinking Paternalism: Power and Parochialism in a Southern Mill Village." *Journal of American History* 83, no. 3 (1996): 872–902.

Winter, Michael. "Embeddedness, the New Food Economy and Defensive Localism." *Journal of Rural Studies* 19, no. 1 (2003): 23–32.

Wirzba, Norman. *The Essential Agrarian Reader: The Future of Culture, Community, and the Land.* Lexington: University Press of Kentucky, 2003.

Wyman, Mark. "Return Migration—Old Story, New Story." *Immigrants and Minorities* 20, no. 1 (2001): 1–18.

Zabin, Carol, Michael Kearney, Anna Garcia, David Runsten, and Carole Nagengast. *Mixtec Migrants in California Agriculture: A New Cycle of Poverty.* Davis: California Institute for Rural Studies, 1993.

Zawisza, Julie. *FDA Announces Findings from Investigation of Foodborne E. Coli O157: H7 Outbreak in Spinach.* Press release. Washington DC: United States Food and Drug Administration, United States Department of Health and Human Services, September 29, 2006. www.fda.gov/NewsEvents/Newsroom /PressAnnouncements/2006/ucm108748.htm.

Zimmerman, Andrew K. "Nineteenth Century Wheat Production in Four New York State Regions: A Comparative Examination." *Hudson Valley Regional Review* 5, no. 2 (1988): 49–62.

Zúñiga, Victor, and Rubén Hernández-León. "Introduction." In *New Destinations of Mexican Immigration in the United States: Community Formation, Local Responses and Inter-group Relations,* edited by Victor Zúñiga and Rubén Hernández-León, xi–xxix. New York: Russell Sage Foundation, 2005.

———. "A New Destination of an Old Migration: Origins, Trajectories and Labor Market Incorporation of Latinos in Dalton, Georgia." In *Latino Workers in the Contemporary South,* edited by Arthur D. Murphy, Colleen Blanchard, and Jennifer A. Hill, 126–35. Athens, GA: University of Georgia Press, 2001.

———, eds (is it ok to add "eds" or do their names need to be rewritten?). *New Destinations of Mexican Immigration in the United States: Community Formation, Local Responses and Inter-group Relations.* New York: Russell Sage Foundation, 2005.

INDEX

abuses, 5, 6, 7, 9, 43, 119
access, job. *See* hiring practices
activism, 3, 7, 19–20, 129–32, 146–50, 152, 156n16
Adams Fairacre Farms, 17
adjusted wages, 51, 116–17
advancement, job, 66–67, 89, 120, 171n69
Adverse Effect Wage Rate (AEWR), 51, 116–17
advocacy, farmworker. *See* farmworker advocates
advocates, farmworker. *See* farmworker advocates
advocates, food movement. *See under* local food movement
advocates, industry. *See* farmers' organizations
AEWR (Adverse Effect Wage Rate), 51, 116–17
African American workers: conflicts with Latino workers, 123; demands, 121; discrimination against, 116–18, 126, 175n2; economic conditions, 36, 127; ethnic succession, 35, 102–3, 105–7, 109, 111–17, 123, 175n2; farm employment decline, 5, 13, 35, 43, 87, 102–3, 109, 110–14, 116–17, 121, 127, 155n7, 177n17; Great Migration, 32, 163n92; labor law politics, 49, 103; migrant stream, 34–35, 111–12, 164n104, 170n57; paternalism, 54; perceptions of, 105, 106–7, 111, 116–18, 121, 123, 124; political influence, lack of, 62, 120–21, 127, 170n57; racism against, 13,

29, 34, 103, 106–7, 116–17, 121, 123, 124, 127, 164n106; stereotypes, 123–24; treatment of, 120–21, 163n92, 170n57; urban migration, 35; welfare, 116, 121, 122, 124; World War II labor shortage, 33, 106, 111–12; youth, 35, 111, 121, 164n109, 177n16. *See also* southern migrants
"aged out" workers, 87, 106–7, 121, 122
Agrarian Dreams (Guthman), 14, 157n17
agrarianism: aesthetics, 41; agriculture industry, 22; beliefs, 21–23, 26, 41, 74; benefits, 172n11; classical writers, 21, 159n20; components, 21; ends, 22, 26; environmental movement, 15, 22; food culture, 2, 10, 16, 21; forms, 159n21; history, 12, 16, 21, 159n20; Hudson Valley, 15–40; ideology, 12–13, 21–23, 26, 74–76, 132, 172n11; Jefferson, 21, 80, 159n20; labor management practices, 10, 12–13; local food movement, 2, 10, 74, 131, 141; marketing, 16, 20, 25–26, 141; myths, 6, 11, 14, 21–22; new forms, 6, 16, 20, 25, 74; organic movement, 25; politics, 21–23, 26, 129, 141, 159n20, 160n37; public perceptions, 2–4, 6–7, 10, 15–16, 20–23, 26, 131, 141, 159n28, 172n11; romanticism, 2, 10, 21, 23, 25, 73, 132, 159n28, 160n37, 172n13; roots, 25, 159n20; self-sufficiency, 2, 21, 74, 172n11; small-scale farming, 10, 21; social virtues, 12, 131; and subsistence growers, 22, 74; sustainability, 22; values, 2, 10, 21–22, 26, 73–76, 160n37, 172n11

nalism, 54, 58–61, 142; promise of, 54, 59, 61, 142

Berry, Wendell, 22–23, 157n16

berry industry: farms, 31, 145; national ranking, 165n124; workers, 32, 42, 62, 106

big-box stores, 17, 38, 70

biodiversity, 20

"birds of passage," 171n68

Bittman, Mark, 131

blacklisting, 52, 109, 119

black workers, 29, 32, 34, 35, 49, 54, 87, 102–27; discrimination, 116–18, 126, 175n2. *See also* African American workers; Caribbean workers

Blue Hill restaurants, 15, 19

Boquerón, Mexico, 178n31

border control, 61, 90–91, 92, 104, 171n67

boutique products, 17, 72, 121. *See also* artisanal foods; niche marketing; value-added products

boycotts, consumer, 134, 147, 149

Bracero program, 112, 168n27

branded products, 15, 157n1

British West Indies program, 112. *See also* H-2 visa program

"buy local" slogan, 2, 76, 148. *See also* local food movement

California, 10, 14, 26, 32; anti-immigration policies, 104; comparison with, 38, 39*table*1, 40, 77; ethical farm model, 145; grape boycott, 134, 147; harvesting methods, 55; immigrant labor in, 104, 124; labor laws in, 134, 167n17; unionization in, 53; 134, 168n31

campaigns, legislative, 22, 131–33, 137–53

Canadian workers, 32, 33, 34

capitalist-industrial food system. *See* industrial food system

carbon footprint, 75, 173n20

career advancement. *See* job advancement

Caribbean workers, 5, 13, 33–35, 43, 102, 107–8, 111, 113–14, 168n27; discrimination, 126. *See also* Haitian workers; Jamaican workers; Puerto Rican workers; West Indies workers

Casa Mono restaurant, 15

cash crop transition, 28

census data, 24, 29, 39*table*1, 78, 136, 144, 151, 161n56, 162n63, 162n75, 169n42, 171n3

Central American produce, 83

Central American workers, 36, 65, 83, 113, 168n27. *See also* Latino workers

Centro Independiente de Trabajadores Agrícolas (CITA), 137, 139, 152

Chávez, César, 134

cheap food policy, U.S., 3, 71, 78–79, 83

cheap labor supply, 3, 12, 23, 27, 30, 36, 107, 120, 127

cheeses, 17, 19, 76, 79

chefs, 15, 18, 19, 72

chemicals, farm, 41, 76, 143, 166n1, 167n18, 171n2, 173n30. *See also* pesticides

Chez Panisse Foundation, 19

child labor, 5, 8, 34, 48, 49, 162n63, 167n18. *See also* youth workers

children, 5, 8, 35, 46, 48, 49, 54, 57, 58, 62–66, 85, 102, 162n63, 167n18

China, 72, 75, 79, 158n2

Chinese workers, 124, 164n98

Church, Frederic Edwin, 19

cider production, 17, 28, 72

citizenship, 26, 55, 67, 93, 114, 127, 170n57; and marriage, 35, 107, 111, 112, 114; and wage hierarchy, 117. *See also* amnesty programs

civic engagement, 11, 20, 26

Civil War, 30, 34

Claiborne, Craig, 18

climate changes, 37, 166n10

CMP (Cornell Migrant Program), 133, 138–39, 181nn25,26,27

Coalition of Immokalee Workers, 53, 134, 144, 146. *See also* Estabrook, Barry; *Tomatoland*

Cohn Farm, 132. *See also* Cornell Migrant Program

Cole, Thomas, 19

collective bargaining: exemption, 6, 23, 49, 53, 131, 141, 155n8, 166n17; legislative action, 131–33, 135, 142; opposition to, 56–57, 132–33, 136, 140, 167n19, 168n27. *See also* farmworker advocates; Farmworkers Fair Labor Practices Act

farmworkers *(continued)*
tionships with farmer, 82, 126, 142, 150; respect, lack of, 44; segregation of, 106, 126; self-empowerment, 112–13; self-perceptions, 44, 170n63; solidarity of, 106, 134; tasks, 42; transportation difficulties, 50, 81, 134; treatment, 44, 163n92, 170n63; white workers, comparisons to, 46, 124; wives of, 57, 58, 65, 85; women, 87, 100*fig*.8, 110, 155n10, 163n89; vulnerability, 5, 42, 48–63, 95–96, 135 170n63. *See also* guest workers; hired hands; housing; labor camps; working conditions; youth workers; *specific ethnic groups*
Farmworkers Fair Labor Practices Act of New York, 131–43, 148, 179n5, 181n32
farm work force: 3, 12, 174n45; cycle of inequality, 67; development services, 86; diversity of, 62, 106, 113; domestic, 88, 121; ethnicity of, 36, 90, 102–27; government regulations, 86; job development services, 86; segregation of, 106; solidarity within, 106. *See also* apprentices; labor; local workers; recruitment
federal laws, 23, 32, 48–49, 138, 155n8, 180n14
federal protections, 49, 51, 137, 155n8
federal regulators, 86
federal requirements, 115, 118, 180n14
field research techniques, 5, 7, 12, 43, 151–53
field workers: daily life, 6, 8, 43–49, 97*fig*.1,2, 98*fig*.3,4, 123, 170n63; disparity from factory workers, 53; sanitation, 86, 136; wages, 51, 63, 163n89; women, 182n37
Filipino workers, 113
Finger Lakes, NY, 111
fired workers. *See* dismissal
first-generation immigrant workers, 31, 93
Fishkill, NY, 28
FLOC (Farm Labor Organizing Committee), 134
floriculture, 37, 71; national ranking, 165n124
Florida, 4, 35, 39*table*1, 51, 53, 62, 105, 106, 107, 108, 113, 134, 146, 164n105, 175n4
foie gras workers, 50, 168n23

food, as common good, 22, 131
food activists, 19, 146–48, 156n16
food advocacy, 1–6, 19, 41, 143. *See also* food activists
food costs, 2–3, 13, 23–24, 30, 38, 49, 78–79, 83, 144–45, 149
food culture: defined 16–17, 20, 158n2; promotion of Hudson Valley, 12, 15–16, 24, 76, 149. *See also* local food movement
food ethic: comprehensive model, 3–4, 10, 12, 14, 49, 67, 128–50; farm model, 145; and labor standards, 41–42, 49, 129, 148; locavores, 2–4, 7, 41–42, 67, 128–31, 148; and marketing benefits, 77, 149; and sustainable employment, 49, 148–49. *See also* ethical farming; ethical food consumption; ethical production
foodies, 15, 56, 130, 147, 164n97, 171n2
food magazines, 56, 173n27. *See also* publications; farm journals
food mileage, 75, 173n20. *See also* transportation
food movement advocates, 2, 75, 129–30, 156n16, 157n17. *See also* local food movement
food policy, 10, 71, 79, 83, 147, 149
food politics, 3, 6–7, 157n16
food quality, 17, 25, 41, 50, 75, 78, 128, 159n15
food safety. *See* public health issues
food writers: classical writers, 159n20; on community aspects, 19, 64, 130; and farmworker justice, 1–5, 11–12, 25, 41; and food movement, 19, 23, 75, 77, 156n16, 157n17; and paternalism; 56; supporting local farms, 15, 41, 73, 128, 141; and worker profiles, 147
forced labor, 1, 5, 43
foreign competition, 37, 69, 71, 75, 79
freeholders, 21–22, 28
free trade agreements, 104, 176n6
Friedland, William H., 170n57
fruit industry, 34, 37, 69; harvesting, 33, 55, 74, 166n7; national ranking, 38, 165n124; production, 2, 6, 11, 15, 40, 47, 73, 79, 108; truck farming, 31–32, 71; workers, 1, 6, 29, 30, 33, 42, 63, 80–81,

86, 95, 109, 112, 166n7. *See also* apple
industry; berry industry
fuel, 38, 72, 83, 130

Garcia, Maria, 1
García-Colón, Ismael, 113
geographic isolation, 12, 48, 50–51, 57, 59,
62–65, 92
George-Warren, Holly, 11
Georgia, 104, 113, 119, 174n45
German immigrants, 28, 29
German prisoners of war, 33, 164n99
Gigi Trattoria restaurant, 15, 18
global food system, 14, 20, 79–80
Glynwood Institute for Sustainable Food
and Farming, 19, 157n1
government: action, need for, 142–45;
agricultural exceptionalism, 22–25,
48–49, 83, 131–33, 160n33, 166n17,
167n21; and agricultural support, 23–25,
79, 80, 131; as labor recruiter, 13, 32, 33,
112, 164n96, 164n98, 164n110; as labor
supplier, 32, 164n100; and land use
initiatives, 165n120; and organic standards, 2; policies, 2, 23, 79, 80; regulations,
86; subsidizing factory farms, 79
grain, 28, 29, 30, 38, 140
grapes, 31, 38, 79; boycott in California, 134,
147; national ranking, 38, 165n124. *See
also* vineyards
Gray, Margaret, 166n16
Great Depression, 32, 40
Great Migration, 32
green cards, 35, 43, 54; guest workers and,
107–9; help obtaining, 59, 67, 85, 91,
102–3, 109, 111, 117–18; statistics, 62, 103;
and wage hierarchy, 117. *See also* guest
workers; noncitizens; undocumented
workers; visas
Greene County, NY, 4*map1*, 39*table1*
"greenhorns," 72, 74, 172n10, 172n13
Greenhorns, The, 74, 172n10
greenhouses, 71
grievances, worker, 108, 132, 142
Griffith, David, 114, 119, 123, 177n19
grocery food markets, 17, 38, 78, 150
growers. *See* farmers
Guatemalan workers, 83, 93, 106, 124

Guerrero, Mexico, 114
guest workers, 3, 12, 33, 35, 46, 91; and
background checks, 178n38; and blacklisting, 119; creation of program, 112,
120, 168n27, 177n23; and ethnic succession, 5–6, 102, 106–9, 111–12, 114–16,
177n23; governmental oversight, 86,
178n38; as labor control, 116; and labor
shortages, 106, 112; organizing efforts,
134; procurement of, 108–9; SAW
provision and, 114; settled-out, 112, 114;
statistics, 3, 12, 111–12, 116; treatment of,
6, 12, 51–53, 58, 119. *See also* Bracero
program; British West Indies Program;
H-2 visa program; *specific ethnic groups*
Gussow, Joan Dye, 78
Guthman, Julie, 14, 25, 157n17

H-2 visa program, 51–53, 91, 109, 112,
168n27, 177n23. *See also* guest workers
Hahamovitch, Cindy, 177n19, 177n23
Haitian workers, 35, 107–8, 111, 113
Hanson, Victor Davis, 160n37
harvest, 29, 33
harvesting, 43, 45, 55, 82, 87, 97*fig.*1,
98*fig.*3,4, 106, 112, 116, 166n7
Hawaii, 166n17
Hawaii Employment Relations Act of 1945,
166n17
hay and silage, 37, 165n124
health insurance, 6, 46, 145
health issues, farmworker, 5, 9, 46, 52–53,
99*fig.*6
heirloom/heritage produce, 15, 17
Herbert, Bob, 50
Hidalgo, Mexico, 110
high-density farming, 72
High Falls, NY, 18
Hindu workers, 124
hired hands, 2–3, 27–28, 29, 36, 161n56,
162n75. *See also* farmworkers
hiring practices, 60, 103, 106–10, 116, 118–
19, 120, 127, 175n5, 176n11
history, farm labor. *See* farmworker history
hoboes, 31, 123
Hofstadter, Richard, 172n11; quoted, 22
home countries: benefits to, 119; blacklisting in, 119; as comparison, 47, 65, 89, 92,

organics *(continued)*
    social justice proxy, 25, 41; standards, 2,
    14, 25, 41, 76; and undocumented
    workers, 121; work force, 41, 84, 89, 121,
    157; WWOOF, 89
Origins of the New South (Woodward),
    181n33
OSHA (Occupational Safety and Health
    Administration) laws, 180n14
out-of-state: farms, 32; workers, 115. *See also*
    specific regions; specific states
outreach programs, 133, 135, 138
overseas competition, 13, 37, 69, 71, 72, 75,
    79
overtime pay: agricultural effects of, 140;
    examples, 49–50, 166n14; lack of, 5, 6, 7,
    23, 84, 120, 122, 131, 166n14, 166n17;
    legislative action and, 131–32, 141, 142

Paarlberg, Don, 23
PACE (Purchase of Agricultural Conserva-
    tion Easements), 37. *See also* land use
packinghouse workers, 43, 45, 100*fig*.7,8,
    106, 110–11
padrone, 33, 163n85
participant observation, 7, 152–53
paternalism, 5, 6, 13, 42, 53–61, 96, 142, 147,
    150, 169n47
pay rate, 29, 50
PDR (Purchase of Development Rights),
    37, 165n120. *See also* land use
pea pickers, 31
Pennsylvania, 31, 33, 34, 38, 165n10
Pennsylvania workers, 31, 33, 34
Pensiero, Laura, 18
periodicals. *See* publications
Perkins, Bill, 139, 181n29
Peruvian workers, 114
pesticides: farmworker health and, 5, 46;
    foreign products, 75, 79; government
    regulations, 76, 80–81, 133, 143; legisla-
    tive efforts, 133; and marketing, 41, 129,
    145, 166n1; organic produce, 41; use, 69,
    74, 80–81, 129, 171n2. *See also* chemicals
physical demands, 1, 6, 8, 87–88, 98*fig*.3,4,
    99*fig*.5,6, 122
pick-your-own farms, 1, 15, 77
"pioneer" workers, 110, 176n8

Pirog, Rich, 173n20
plantation system, 27, 54
planting, 6, 42–43, 88, 166n7
plight: of farmers, 68, 140–41; of workers,
    30, 50, 64, 95–96, 128, 168n23, 181n33
Polish immigrants, 33
political action, 57, 130–31
political influence, lack of: 62, 120–21, 127,
    170n57
political system: Democrats, 49, 135, 140,
    182n34; effects on food prices, 79 farm-
    ers' organizations, 49; federal, 131;
    politicians, 7, 41, 176n10; power, lack of,
    49, 55, 62; Republicans, 135, 140; south-
    ern legislators, 167n21; state, 131–32, 152,
    180n12; white supremacists, 49
Pollan, Michael, 41, 77, 80, 128, 130–31
Portes, Alejandro, 174n55
Portuguese immigrants, 124
potatoes, 65, 70, 100*fig*.7, 111
Poughkeepsie, NY, 28, 29, 114, 178n31
poultry industry, 28, 43, 50, 69, 157n20
poverty, 96, 148; cyclical migrations,
    93–94; institutionalized, 67; level, 44,
    65, 166n11; social marginalization, 48,
    55, 57; reports, 146. *See also*
    impoverishment
Power to Choose, The (Kabeer), 155n10
prejudice, 33, 164n96, 164n99
"price of proximity," 54, 56–57, 142, 149
prices. *See* food costs
primary sector, 89, 176n7. *See also* dual
    labor market theory; secondary sector
prison camps, 33, 164n96
prisoner of war labor, 33, 164n96
private sector, 155
produce, local: consumer demand, 10–11,
    78; direct marketing, 72–73, 77, 130,
    144–45; diversity, 72–73; farmworker
    perk, 58, 89; large retailers, 38, 77–79;
    pesticide use, 171n2; public awareness,
    1–2, 15–19, 78, 121
product diversity, 20, 31, 41, 55, 72–73, 129,
    140
profiling, 29, 123–26
profitability: factors impacting, 36, 44, 69,
    71; lack of, 75; statistics, 16, 77–78,
    171n3